Sermon On The Beach

Jeremy Griffith

Watch the video of this presentation at
www.HumanCondition.com/
sermon-on-the-beach

OR

Scan code to view

Sermon On The Beach by Jeremy Griffith

Published in 2024 by WTM Publishing and Communications Pty Ltd
(ACN 103 136 778) (www.wtmpublishing.com).

All enquiries to:

WORLD TRANSFORMATION MOVEMENT®
Email: info@worldtransformation.com
Website: www.humancondition.com or www.worldtransformation.com

The World Transformation Movement (WTM) is a global not-for-profit movement
represented by WTM charities and centres around the world.

ISBN 978-1-74129-100-1
CIP – Biology, Philosophy, Psychology, Health

Commendations for Griffith's treatise

From Thought Leaders

'[**Prof. Stephen Hawking**] is most interested in your impressive proposal.'
● 'In all of written history there are only 2 or 3 people who've been able to think on this scale about the human condition.' **Prof. Anthony Barnett**, zoologist ● '*FREEDOM* is the book that saves the world...cometh the hour, cometh the man.' **Prof. Harry Prosen**, former Pres. Canadian Psychiatric Assn. ● 'I am stunned and honored to have lived to see the coming of "Darwin II".' **Prof. Stuart Hurlbert**, esteemed ecologist ● 'Living without this understanding is like living back in the stone age, that's how massive the change it brings is!' **Prof. Karen Riley**, clinical pharmacist ● 'Frankly, I am blown away by the ground-breaking significance of this work.' **Prof. Patricia Glazebrook**, philosopher ● 'I've no doubt a fascinating television series could be made based upon this.' **Sir David Attenborough** ● '*FREEDOM* is the necessary breakthrough in the critical issue of needing to understand ourselves.' **Prof. David J. Chivers**, former Pres. Primate Society of Britain ● 'Whack! Wham! I was converted by Griffith's erudite explanation for our behaviour.' **Macushla O'Loan**, *Executive Women's Report* ● 'This is indeed impressive.' **Dr Roger Lewin**, preeminent science writer ● 'I have recommended Griffith's work for his razor-sharp biological clarifications.' **Prof. Scott Churchill**, psychologist ● 'An original and inspiring understanding of us.' **Prof. Charles Birch**, zoologist ● 'The insights are fascinating and pertinent and must be disseminated.' **Dr George Schaller**, preeminent biologist ● 'Very impressive, particularly liked the primatology section.' **Prof. Stephen Oppenheimer**, geneticist, author *Out of Eden* ● 'I consider the book to be the work of a prophet.' **Dr Ron Strahan**, former dir. Sydney Taronga Zoo ● 'The scholarly value [of Griffith's synthesis] is comparable to several of the most celebrated publications in biology.' **Prof. Walter Hartwig**, anthropologist ● 'I believe you're on to getting answers to much that has bewildered humans.' **Dr Ian Player**, famous Sth. Afr. conservationist ● 'A superb book, a forward view of a world of humans no longer in naked competition.' **Prof. John Morton**, zoologist ● 'This might bring about a paradigm shift in the self-image of humanity.' **Prof. Mihaly Csikszentmihalyi**, psychologist ● 'As a therapist this is a simply brilliant explanation.' **Jayson Firmager**, founder of *Holistic Therapist Magazine* ● 'The questions you raise stagger me into silence; most admirable.' **Ian Frazier**, author *Great Plains* bestseller ● 'The WTM is an island of sanity in a sea of madness.' **Tim Macartney-Snape**, world-leading mountaineer & twice Order of Australia recipient

Commendations From The General Public

'Griffith should be given Nobel prizes for peace, biology, medicine; actually every Nobel prize there is!' ● 'He nailed it, nailed the whole thing, just like the world going from FLAT to ROUND, BOOM the WHOLE WORLD CHANGES, no joke.' ● '*FREEDOM* will be the most influential, world-changing book in history, and time will now be delineated as BG, before Griffith, or AG, after Griffith.' ● 'I'm speechless – this is bigger than natural selection & the theory of relativity!' ● 'I really think this man will become recognized as the best thinker this world's ever seen, and don't we need him right now!' ● 'Griffith has decoded the human species, we FINALLY know what's going on & the suffering stops!' ● 'The world can't deny this for much longer, let the light in, save the human race!' ● 'This is the most exciting moment in my life. *THE Interview* tore my hat off & let my brain fly into the sky!' ● '*THE Interview* should be globally broadcast daily. The healing explanation humans so sorely need.' ● 'In a world that's lost its way there's no greater breakthrough, water to a world dying of thirst.' ● 'Dawn has come at Midnight! A brilliant exposition, we could be on the cusp of regaining Paradise!' ● 'This man has broken the great silence, defeated our denial, got the truth up, woken us from a great trance.' ● 'Beware the 'deaf effect; your mind will initially resist the issue of our corrupted condition and so find it hard to take in or hear what's being said, but if you're patient you'll find the redeeming explanation of our condition pure relief.' ● 'John Lennon pleaded "just give me some truth", well this site finally gives us *all* the truth!' ● '*FREEDOM* is the most profound book since the Bible, now with the redeeming truth about us humans.' ● '*Death by Dogma* is brilliant clarification.' ● 'We were given a computer brain, but no program for it; but Aha, Griffith has found it, made sense of our lives!' ● 'This just goes deeper & deeper in explaining us, like dawn devouring darkness, amazing!' ● 'Agree, this is not another deluded, pseudo idealistic, PC, 'woke', false start to a better world, but the human-condition-resolved real solution.' ● 'Freedom indeed! What we have here is the second coming of innocence who exposes us but sets us free!' ● 'As prophesised, King Arthur has returned to save us (mentioned in par.1036 *Freedom*)' ● 'We all need to go back to school & learn this truthful explanation of life.' ● 'Join in our jubilation, your magic reunites, all men become brothers, all good all bad, be embraced millions! This kiss [of understanding] for the whole world' – From Beethoven's 9th (par.1049 *Freedom*)

Contents

Background

Jeremy Griffith is an Australian biologist who has dedicated his life to bringing redeeming and psychologically healing biological understanding to the dilemma of the human condition—which is the underlying issue in all human life of our species' extraordinary capacity for what has been called 'good' and 'evil'.

Jeremy has published over ten books on the human condition, including:

— *Beyond The Human Condition* (1991), his widely acclaimed second book;

— *A Species In Denial* (2003), an Australasian bestseller;

— *FREEDOM: The End Of The Human Condition* (2016), his definitive treatise;

— *THE Interview* (2020), the transcript of acclaimed British actor and broadcaster Craig Conway's world-changing and world-saving interview with Jeremy about his book *FREEDOM*;

— *Death by Dogma: The biological reason why the Left is leading us to extinction, and the solution* (2021), which presents the biological reason why Critical Theory threatens to destroy the human race;

— *The Great Guilt that causes the Deaf Effect* (2022), which describes how lifting the great burden of guilt from the human race initially causes a 'Deaf Effect' difficulty taking in or 'hearing' what's being presented;

— *The Shock Of Change that understanding the human condition brings* (2022), which addresses how to manage the shock of change that inevitably occurs when the redeeming understanding of our corrupted condition arrives;

— *Therapy For The Human Condition* (2023), which is about the therapy that is desperately needed to rehabilitate the human race from our psychologically upset state or condition, elaborating on what is presented in *FREEDOM*;

— *Our Meaning* (2023), which explains how being able to know and fulfil the great objective and meaning of human existence finally ends human suffering;

— *The Great Transformation: How understanding the human condition actually transforms the human race* (2023), which gives a concise description of how the psychological rehabilitation of humans occurs, and how everyone's life can immediately be transformed; and

— *AI, Aliens & Conspiracies: The Truthful Analysis* (2023), which provides Jeremy's thoughts on the much discussed question of the danger of Artificial Intelligence (AI), and on the possibility of alien life visiting Earth, and also his explanation for the epidemic of conspiracy theories.

This booklet, ***Sermon On The Beach***, is Jeremy's elaborated transcript of his inspired description of how the human race now leaves the horror of the human condition forever! The video can be viewed at www.humancondition.com/sermon-on-the-beach.

Jeremy's work has attracted the support of such eminent scientists as the former President of the Canadian Psychiatric Association Professor Harry Prosen, the esteemed American ecologist Professor Stuart Hurlbert, Australia's Templeton Prize-winning biologist Professor Charles Birch, the former President of the Primate Society of Great Britain Professor David Chivers, Nobel Prize-winning physicist Stephen Hawking, as well as other distinguished thinkers such as the pre-eminent philosopher Sir Laurens van der Post.

Jeremy is the founder and a patron of the World Transformation Movement (WTM)—see www.HumanCondition.com.

With the real problem of the human condition finally solved we can now ACTUALLY fix the world!

Sermon On The Beach

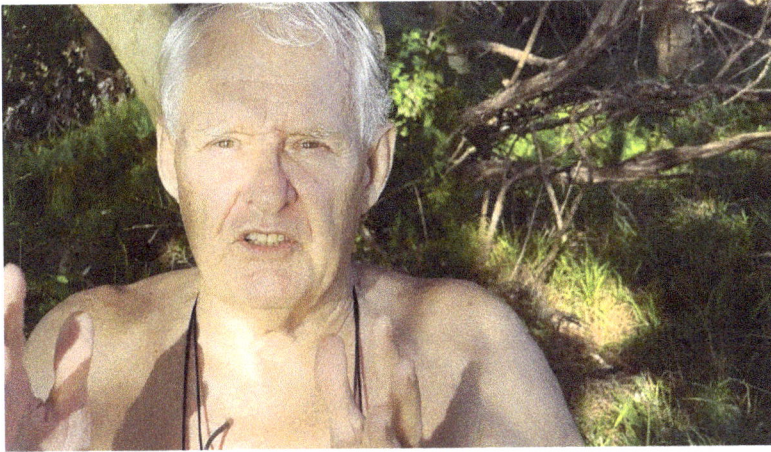

Jeremy Griffith presenting his *Sermon On The Beach*, Sydney, Australia, 2024

[1] **Jeremy Griffith**: [Preview extract from the talk]: "We go from a state of depression about the state of the world and our own lives, to a state of incredible excitement and happiness. And once you get a taste, just a glimmer of experience of what it's like to live free of the agony of the human condition—for how exciting that will be—then you've finally got a taste of how this all works."

---

[2] **Introduction by Tony Gowing** (WTM founding member from the Sydney WTM Centre): This 2024 video presents biologist Jeremy Griffith's already famous summarising 'Sermon On The Beach' (as people are calling it) SOLUTION TO ALL YOUR, AND THE HUMAN RACE'S, PROBLEMS! Now, this presentation assumes you have

already at least watched or read *THE Interview* with Jeremy that appears at the top of our homepage at HumanCondition.com, because it will significantly help you understand the basics of what Jeremy is talking about, so we encourage you to do that if you haven't already done so.

[3] Firstly, people aren't calling it a 'Sermon' because it's religious, but because, like Christ's 'Sermon on the Mount'—which was his great call to action, his great selling of the wonders of Christianity's ability to *relieve* humans of their distressed human condition—Jeremy's presentation is his great call to action to actually *end* the agony of the human condition that the finding of the redeeming and rehabilitating biological explanation of the human condition finally makes possible.

[4] And what Jeremy is 'summarising' is how this great rehabilitation of the human race takes place, which is presented in more detail in *The Great Transformation* video and booklet. Both it and this presentation are freely available at HumanCondition.com.

[5] The reason this talk by Jeremy has already become famous is because, as a viewer said, 'It absolutely floodlights the pathway that the whole human race now takes out of Plato's dark cave of denial and delusion that has been crippling the human race and which we have all had enough of!' Yes, what this truly incredible talk does is cement in the foundations for what will become the biggest and most exciting movement the world has ever seen—because it is the realisation of what the scientist-philosopher Teilhard de Chardin anticipated when he wrote, 'The Truth has to only appear once...for it to be impossible for anything ever to prevent it from spreading universally and setting everything ablaze'!

[6] I also want to strongly recommend reading the transcript of this talk because Jeremy has added many extremely helpful and enlightening elaborations to what he says in this incredible talk.

---

[7]**Jeremy Griffith**: [Further preview extract from the talk: "Honestly, it'll flip your situation from a state of constant ambivalence and struggle and desperation to a state of absolutely amazing excitement!"]

[8]I'm down by the beach sitting under some banksia trees, and I just want to try to capture how the Transformed Way of Living works—try to give as *powerful* a description of how it works as I can possibly think of. And to do that, the idea has occurred to me to ask you to imagine if the situation in the world becomes untenable—because it *is* daily becoming more desperate, everyone's becoming more and more concerned about the plight of the human race. There's desperation and psychosis everywhere, and conflict and destitution and devastation. So it's actually on everyone's mind now that it's become undeniable that we've got to do something fundamental about the plight of the world, the plight of our species.

[9]As I have explained in all my presentations (such as in paragraphs 37-38 of *THE Interview*, and pars 70-100 of *The Great Transformation*), denial of the truth of our species' original innocent, cooperative, selfless and loving instinctive self or soul and of our present competitive, selfish and aggressive corruption of it is how we've had to cope with the shame and guilt of that corrupted state while we couldn't explain it. Any way we could, we had to try to maintain a positive view and presentation of ourselves in order to hold back the unbearably depressing truth of our horrifically soul-corrupted so-called 'human condition'. So we build this 'castle' of trophies—we become the captain of the football club, or successful at work, etc, etc—so that we can feel at least a little bit good, worthy and positive about ourselves; a castle of wins that maintain our sense of worth against the unbearable truth of our species' horrific 2-million-year long, conscious-mind-induced, corrupted condition—which our encounter with when we were growing up inevitably led to our own horrific corrupted condition!

[10] Not thinking too deeply or honestly, and staying positive, have been our ways of holding back the unbearably—in fact, suicidally—depressing truth of our corrupted condition that we have never been able—until now—to explain and by so doing end the guilt and shame of. This is a description of the excruciating depression that the philosopher <u>René Descartes</u> felt when he tried to confront the horror of his corrupted condition: **'So serious are the doubts into which I have been thrown…that I can neither put them out of my mind nor see any way of resolving them. It feels as if I have fallen unexpectedly into a deep whirlpool which tumbles me around so that I can neither stand on the bottom nor swim up to the top'** (*Second Meditation*, 1641; tr. J. Cottingham, 1984). Tragically, unbearably, as the psychoanalyst <u>Carl Jung</u> said, **'When it [our shadow] appears…it is quite within the bounds of possibility for a man to recognize the relative evil of his nature, but it is a rare and shattering experience for him to gaze into the face of absolute evil'** (*Aion: Researches into the Phenomenology of the Self*, 1959; tr. R. Hull, *The Collected Works of C.G. Jung*, Vol. 9/2, p.10). Yes, while we couldn't explain it, it *has been* a **'shattering experience'** seeing the total contrast between—**'the relative evil'** of—our present angry, egocentric and alienated condition and our species' original all-loving and all-sensitive existence, which our instinctive self or soul—our conscience—is aware of and expects us to behave in accordance with!

[11] <u>So finding any way we could to avoid the **'shattering experience'** of self-confrontation with the issue of our corrupted condition, and feel at least a little bit worthy and good and honourable has been the great preoccupation of our lives—which is why our 'trophy room' of wins, our 'castle' of successes have become so incredibly precious to us!</u>

[12] <u>Yes those successes and reinforcements, those artificial ways of maintaining our sense of worth, have been *immensely* important to us, BUT, AS I NOW WANT TO EXPLAIN, THE PROBLEM IS our extreme attachment to this *artificial* defensive structure makes the idea that we should give it up, now that we have found the *real* defence of our corrupted condition of the biological explanation of it, almost impossible to even consider, let alone accept!</u>

[13] The fact is that blocking out the truth of our corrupted condition and maintaining our artificial ways of feeling good about ourselves have been so immensely important to us that we don't even admit there is anything fundamentally wrong with us or with our existence; we have coped with the truth of our species', and of our own, corrupted, soul-destroyed condition by completely denying that it exists: 'What do you mean the guilt and shame of my and the human race's corrupted condition? I don't suffer from some corrupted condition, it doesn't exist, there is nothing basically wrong with me or my life!' That is how fearful of and insecure about our corrupted, soul-devastated condition we have been—we haven't even been able to admit it exists!

[14] But our world is becoming so desperately messed up by our corrupted angry, egocentric and alienated, competitive, selfish and aggressive human condition that we can no longer pretend that everything is basically fine and we are okay. In fact, it's become undeniable that we're heading for some sort of catastrophe, some sort of end play situation where the human race and our planet can't endure any more of our psychologically upset, soul-corrupted, competitive, selfish and aggressive human condition. As Professor Harry Prosen, a former President of the Canadian Psychiatric Association and Patron of the World Transformation Movement (WTM) that promotes my work, wrote, **'I think the fastest growing realization everywhere is that humanity can't go on the way it is going.'** So, yes, across the world there's a sort of waking up to the fact that enough is enough, we need some fundamental solution to the plight of the human race, to the plight of our world.

[15] So despite our practice of denying our corrupted condition and artificially maintaining a positive view of ourselves and our way of living, it has become undeniable that there is some sort of deep, universal psychosis troubling the human race that we have to fix. In effect we are being forced to face the issue of our corrupted human condition, which, despite our historic denial of that condition, should make us more receptive to the description, analysis and explanation

of that condition that I present in my work—and that is starting to happen, as I will illustrate later. THE PROBLEM, HOWEVER, THAT THEN OCCURS, which is what I want to talk about now, is that this biological explanation of the human condition that explains why we humans are good and not bad clearly obsoletes the need for our present artificial angry, egocentric and alienated ways of coping with the human condition and allows us to adopt the Transformed Way of Living that is described at length in *The Great Transformation* video and booklet, where we live in support of that redeeming understanding of ourselves. With the real defence of our corrupted condition found of the redeeming explanation for it, we no longer need to use our defensive angry, egocentric and alienated ways of coping with it, we can let that way of living go, BUT the problem is how do we let go of our attachment to those old defences that we have become so dependent on and practiced at using?

[16]We are caught in a conflict. We have developed this strategy since we were adolescents of sustaining our sense of worth artificially through attacking, denying and proving wrong our apparently indefensibly corrupted condition. In fact this defensive structure that we have built up to maintain our sense of worth is *so* important to us that we spend every minute of every day preoccupied in our mind maintaining it, keeping at bay any chance of connecting with the **'shattering experience'** of the seeming **'evil'** of our nature, as Jung described our situation. So it's a huge industry that has owned our mind—again, because it is all we have had to sustain us, it's what has kept us going, it's what has allowed us to cope with the unbearable truth of our corrupted condition. So this idea that an alternative way of living is now possible through the finding of the redeeming understanding of our corrupted human condition, where we can let go of our carefully constructed and maintained artificial defences of ourselves like winning power and glory and by so doing become transformed from living so defensively, deludedly and dishonestly, runs in the face of this immense historic commitment we have to sustaining ourselves, such as through winning as many accolades

and getting as much artificial and superficial reinforcement as we possibly can. Lacking the *real* defence of the 'instinct versus intellect' explanation for our corrupted condition—which we have now finally found—all we have had to cope with the unbearable truth of our corrupted condition are the *artificial* defences of attacking any criticism of ourselves, of denying and blocking out any such criticism, and of achieving any positive reinforcement of ourselves that we can get hold of; our angry, alienated and egocentric ways of coping are what have sustained us, and as a result we can't easily let them go.

[17] As I said, every second of every day our mind has been preoccupied maintaining and building these artificial ways of sustaining our sense of self-worth. So it's a very, *very* powerful attachment we have to that artificial defensive structure. And now this explanation of the human condition comes along that explains biologically why we humans are good and not bad, and by so doing obsoletes the need for our artificial angry, egocentric and alienated ways of living, and allows us to adopt the Transformed Way of Living where we live in support of that redeeming understanding of ourselves. But again, if you stand back and think about that situation, there is a problem with it. Our attachment to all our denials of having a corrupted soul, and all our artificial self-worth-sustaining wins and glorifications of ourselves have been so precious to us and are thus so strong that they are in extreme conflict with this idea that we should let that carefully worked on and maintained structure go!

[18] We are caught in a 'Mexican Standoff' between our attachment to our artificial defensive structure, and understanding that this information defends the human race now from a fundamental, first-principle, biological basis and therefore that we don't need all those artificial defences of attacking, blocking out and chasing wins anymore, and can let them go. We know the human condition has been explained, we can accept that is true and so we don't want to deny it, but it is not at all easy for us to accept the implications of that, which is to let that way of living go. We are in a Mexican Standoff where we can't go back into denial, pretence and delusion and yet we can't go forward

into accepting and implementing the implications of what we have learnt! We are so attached to our artificial defensive structure, our ways of sustaining our sense of worth, that this idea that we should let that go is just an anathema; an 'anathema' being something that's loathsome, extremely unacceptable. So we're caught in this conflict between what's been sustaining us and the logic that says we can— and now must if the human race is to be transformed from the now unendurable effects of that way of living—let all that go!

Our 2-million-year-practiced angry, egocentric and alienated way of living
in denial, pretence and delusion about our corrupted condition

The 'Mexican Standoff' situation where you don't want to deny the truth of our
corrupted human condition but also don't want to let go your now obsoleted
artificial angry, egocentric and alienated ways of sustaining your sense of worth

The Transformed Way of Living where you have let go of your angry,
egocentric and alienated way of defending yourself and are living in
support of the redeeming understanding of our corrupted condition
that enables you to effectively live free of the human condition

[19] So, yes, this idea that we should take up this Transformed Way
of Living and live in support of this all-redeeming, all-reconciling,
all-relieving and thus all-wonderful understanding of our corrupted
condition that completely obsoletes the need for our old—once
necessary but now unnecessary—artificial, angry, egocentric
and alienated ways of coping with our corrupted condition *is an
anathema*; it's a loathsome idea.

[20] Now—and this is very important, in fact it is key—the great
problem with this resistance, this state of being stuck in the Mexican
Standoff, is that IT BLOCKS OUR MIND FROM BEING ABLE TO
EXPERIENCE HOW THIS TRANSFORMED WAY OF LIVING ACTUALLY
WORKS, which I am working my way towards explaining.

[21] So we've got this artificial defensive structure; we man-
ufacture this smiling, positive, 'I'm okay—in fact I'm not just
okay, I'm a bloody legend'-persona, and so on, and so on. But the
reality is that this artificial angry, egocentric and alienated way of
defending ourselves, and the competitive, selfish and aggressive
behaviour it results in, is obviously sooner or later going to become

unsustainable. Competitively, selfishly and aggressively beating each other up, smashing and grabbing whatever wins we can, etc, etc, is ultimately not a viable way to live. And despite our carefully maintained delusions that we and our world are okay, that untenable situation is where the human race has now arrived. The situation where our artificial ways of defending and reinforcing ourselves, our angry, egocentric and alienated ways of coping with the unbearable condemnation we have felt for corrupting our loving soul—which, as I explain in all my publications, has been absolutely necessary because it has enabled us to continue our heroic search for knowledge, ultimately for self-knowledge, the now found redeeming and relieving good reason for why we corrupted our loving soul—leads to more and more psychological upset, more and more conflict and corruption and opportunism and devastation, so it *is* ultimately an unsustainable way of living.

²²For some 2 million years we humans have had to try to maintain a positive view of ourselves and of our way of living; basically maintain that there is nothing fundamentally wrong with our lives—'We don't need any great change to what we're doing, we're doing really well', that sort of thing. So we haven't allowed ourselves to even think about, or envisage, any change to that view. That's why people experience the 'Deaf Effect' when they read or listen to this information—because they are reading or listening to all this stuff about our lost innocence and so forth, all this truth about our corrupted condition, which is totally at odds with the almost universal strategy of completely denying the existence of our corrupted condition and instead maintaining all manner of artificial ways of feeling good about ourselves. Understanding the human condition brings incredible relief and excitement, but with their mind thus blocked from even being able to take in or 'hear' these explanations and descriptions of our corrupted condition, their mind is unable to discover that relief and excitement of having this relieving understanding of ourselves (see an example of

this excitement in Michael Perry's Affirmation at the conclusion of this presentation). As one sufferer of the Deaf Effect described their experience: **'Being in the WTM's Facebook Group I felt like there is a party going on that I haven't been invited to—in fact this was so much the case I ended up feeling there must be something wrong with me!'** Thankfully, through patiently persevering listening to and reading and digesting the presentations about the human condition the Deaf Effect can be overcome, so the 'Digest Effect' was how that sufferer of the Deaf Effect overcame it! (You can learn much more about the Deaf Effect in my video and booklet *The Great Guilt*.)

The initial 'Deaf Effect' reaction to description of the human condition

Persevering enables you to overcome the 'Deaf Effect' and access understanding of every aspect of human existence

[23] So letting that artificial defensive structure go is not something we can easily allow our mind to consider. Once through the Deaf Effect and thus able to take in this biological defence of the human condition, we can follow the logic that having that defence means our old artificial reinforcement system is finished with and we can now live in support of this truthful 'instinct versus intellect' explanation of ourselves instead of in support of our once necessary but now no longer necessary angry, egocentric and alienated way

of coping, and by so doing fix the world. We can tick all those boxes about how true and effective supporting this information would be; we can be happy with the logic of that—but, again, the problem when you get down to it is that that change, that letting go, is in complete conflict with our artificial angry, egocentric and alienated defensive structure, so it's not something our mind can easily consider letting go of.

[24]We can accept all the logic but we're caught in the Mexican Standoff where we struggle to change the focus of our mind from maintaining our artificial defensive structure, and let that focus go and instead focus on supporting the real defence for ourselves of the explanation of our corrupted condition—a corrupted condition that we have been determinedly denying the existence of but can now see as a result of being able to understand the human condition does in fact exist. And so—and this is the very important point I mentioned earlier—because we find it almost impossible changing the focus of our minds, WE ARE UNABLE TO DISCOVER HOW INCREDIBLY RELIEVING AND EXCITING THE NEW FOCUS FOR OUR MINDS IS. Our extreme attachment to our old defensive structure blocks our mind from discovering how fabulously satisfying, wonderful and exciting the new Transformed Way of Living is. In his presentation in *The Great Transformation* video and booklet (which everyone should watch or read before watching or reading this presentation), WTM founding member Tony Gowing spoke of the clash between our old defensive structure and the Transformed Way of Living. He described our extreme attachment to our carefully constructed and constantly maintained denial-practising and power-fame-fortune-and-glory-winning defensive structure that has been incredibly precious to us because it allowed us to live by keeping at bay the suicidal depression that our corrupted condition would otherwise cause us, and the idea that we should let all that go, as being 'like two freight trains coming from opposite directions down a single track in a giant game of chicken!'.

[25] THE VERY BIG QUESTION THEN IS, HOW TO BREAK THIS EXTREME ATTACHMENT WE HAVE TO OUR OLD DEFENSIVE STRUCTURE THAT IS BLOCKING US FROM DISCOVERING HOW FABULOUSLY RELIEVING AND EXCITING THE NEW TRANSFORMED WAY OF LIVING IS—WHICH IS WHAT IS NEEDED TO END OUR EXTREME ATTACHMENT TO THAT OLD DEFENSIVE STRUCTURE? Those suffering from the Mexican Standoff are caught in a Catch-22—you resist taking up the Transformed Way of Living until you discover how good it is, but you don't discover how good it is until you take up the Transformed Way of Living!

[26] Now I suggest the most effective way of overcoming this problem of your extreme attachment to your existing defensive structure and its block to being able to experience how incredibly relieving and exciting the new Transformed Way of Living is, is to think very clearly about the situation the human race is in, which, if you have understood the explanation of the human condition, you can now do. So this is that clear thinking that I am suggesting is required. The truth is that our artificial angry, egocentric and alienated ways of coping with the unbearable truth of the human race's and our own corrupted condition—our artificial ways of maintaining our sense of self-worth—are ultimately not sustainable and therefore it is undeniable that at some stage we have to stop living that way. The other truth, that understanding of our corrupted human condition allows us to now recognise, is that this end point where the human race can't endure any more angry, egocentric and alienated behaviour has now arrived, which means the human race *urgently has to* make the change from living that way and instead take up the Transformed Way of Living, which means you should

make that change yourself, take up the Transformed Way of Living. Basically, you need to get all the logic about the true situation that the human race is in clearly lined up in your mind. Sure, you are hugely attached to your denial-practising and power-and-glory-focused way of living, but that way of living is ultimately not sustainable; so ultimately everyone has to change from living that way, and since the human race *has* actually reached that end point where we have to leave that way of living, and since we *can* now do that by taking up the Transformed Way of Living, taking that up is what *everyone now has to do*—and since it is now a priority that everyone makes that change, *you need to support that change* by making it yourself. Essentially, you have to defeat your attachment to your artificial defensive structure with an extremely clear understanding of the human race's, and of your own, situation. What you are doing is holding all the logic so clearly in your mind that you are able to allow your mind to actually take up the Transformed Way of Living, AND AT THAT MOMENT WHEN YOU ACTUALLY DO MAKE THAT CHANGE YOU WILL SUDDENLY EXPERIENCE HOW INCREDIBLY RELIEVING AND EXCITING LIVING THAT WAY IS, AND THEN YOU WON'T WANT TO LIVE ANY OTHER WAY EVER AGAIN!

[27] Our old defensive structure is clearly built on false foundations. Attacking, trying to prove wrong by winning accolades, and trying to block out from our mind, the implication that we are bad or evil or unworthy for having a corrupted instinctive self or soul are *artificial* ways of coping with the historically unbearable truth of our corrupted condition. They are not the *real* defence of the explanation for why we are not bad or evil or unworthy for having corrupted our all-loving and all-sensitive instinctive self or soul. Further, since they are not the *real* defence for our corrupted condition, they require constant maintenance; we have to keep attacking any perceived criticism, keep chasing reinforcing wins, and keep blocking out any perceived condemnation. So the truth is that it is a very insecure and fragile structure we are having to constantly maintain. As a result of our differing encounters with the upset, corrupted state of the human

condition, everyone (or virtually everyone because there are very rare individuals who have sufficiently escaped encounter with the upset state of the human condition in their upbringing to not have to adopt a human-condition-denying defensive way of living) is variously insecure, uncertain of their worth, which means virtually everyone is variously having to try to maintain their psychological denials, which is their psychosis, and their self-validation 'castle' of reinforcements, the result of which is that there is all this psychosis and all this competition, selfishness and aggression everywhere, which can only become worse and worse if there is no change, so that whole activity *is* ultimately not sustainable. As I relate in *The Great Transformation*, my headmaster at Geelong Grammar School, Sir James Darling recognised this unsustainability in his absolutely astonishingly truthful and thus insightful (he was, as was recognised in his 1995 full-page obituary in *The Australian* newspaper, **'a prophet in the true biblical sense'**) 1950 Speech Day address, saying, **'selfishness is, as it has ever been, the ultimately destructive force in a society, and there are only two cures for selfishness—the regimented state which we all profess to dislike, and the change of heart, which we refuse to make. That is the choice, believe me, for each one of us, and we have not much time in which to make it. The need for decision** [to have a 'change of heart' and live selflessly] **is serious and urgent, and the sands** [of time] **are running out'** (Weston Bate, *Light Blue Down Under*, 1990, p.219 of 386). (See more of Sir James's astonishingly profound thinking in Part 5 of *The Great Transformation*.) Yes, as my explanation of the human condition finally makes understandable, selfishness has been necessary during humanity's heroic search for understanding of its corrupted condition, however it is still, as Sir James said, **'the ultimately destructive force in a society'**. And yes, as I explain in my book *Death By Dogma*, the pseudo idealistic dogma of the political Left's **'regimented state'** is **'dislik**[able]**'**. And yes, as we can now understand, we have had to **'refuse to make' 'the change of heart'** to living selflessly while we still had to persevere with our corrupting search for understanding of our divisive, corrupted, 'fallen' human condition, but now that we have found that understanding we can have that **'cure' 'for selfishness'** of **'the change of heart'** to living selflessly that Sir James emphasised even 74 years ago now back in 1950 was **'serious and urgent'** because **'the sands** [of time] **are running out'.** There simply *has to be* a change from living in ever more insecure states of psychotic denial, and ever more egocentrically competitively and selfishly, to living free of those insecure preoccupations, in psychologically healing honesty (see my book *Therapy For The Human Condition*), and in cooperative and selfless support of the secure, *true* defence for our corrupted condition, which is the human-race-transforming way of living. Okay, but again, how do we overcome our attachment to our psychological denials and our egocentric castle that we've been depending on, and by so doing discover how fabulously relieving, human-race-and-world-fixing, and incredibly exciting it is to live in support of this true defence of our corrupted condition?

[28] What I am suggesting is if you imagine that the world becomes more and more angry, egocentric and psychologically alienated/psychotic (which is what has been happening at a compounding rate), then clearly that artificial way of defending ourselves has to end. The artificial angry, egocentric and alienated ways of defending ourselves that we have been employing where everyone has been practising more and more alienating denial, and more and more aggressively seeking a win (and if they can't get a win they all too often resort to finding illegal, short-cut, corrupt ways to get a win, which leads to the complete breakdown of society, such as is occurring in many countries now, like South Africa) *is* a structure that is ultimately doomed to fail. And then imagine that the human race has arrived at this end point, doomed situation where we humans and our world can't endure any more of this psychologically upset way of defending ourselves that we have had to depend on to artificially maintain our sense of self-worth (which is actually the exhausted state that the human race has arrived at, but which we have not been allowing ourselves to see because we have had to maintain a positive outlook). So, if the human race *has* arrived

at this end point, apocalyptic situation (which, again, it actually has, with younger generations now too psychologically distressed, overwhelmed and crippled to cope anymore, and the old business, political and social systems no longer working), what is going to happen? I'm asking everyone in the Mexican Standoff situation to fully imagine the world getting to that dire end play situation where terminal levels of alienation/psychosis/soul-devastation and greed and hunger for power are going to cause the extinction of the human race, and it is all going to occur in the most horrifically agonising, paralysed-with-anxiety-and-depression, crippled-to-the-core, way (which, again, is a situation we have actually arrived at, as the epidemics that are occurring of extreme psychoses like ADHD and autism show—it's said that the outflow of water to the sea from cities in California is now so full of anti-depressant medication that it's toxic to marine life! Read more about this end play situation the world is at in Freedom Essay 55). And then imagine at that point that the understanding of the human condition arrives that obsoletes the need for our old artificial, denial-practising, competitive, selfish and aggressive, power, fame, fortune and glory, defensive ways of living, and we can live with truth and honesty, cooperatively and selflessly free of all that preoccupation in support of the redeeming, reconciling and healing understanding of our corrupted human condition (which is the key breakthrough that actually has arrived), which is a change that will totally fix the world. Suddenly there is a solution to all of our and the world's excruciating problems in front of us that is 100% effective and 100% fixes everything (which it actually does). And so we've got a polarised situation where we know the world is dreadfully, horrifically messed up, but we are hugely attached to our artificial defence structure that we've built and are manicuring every day, but it has run its course, it's had its day, and so we've hit end point where that way of living is no longer sustainable. And so, I want everyone caught between the old way of living and the new way to imagine that they have arrived at an acceptance of this situation in their mind where the predicament of the human race really *is* desperate and we *absolutely need* a solution. And on the other side of this polarised situation where we are hugely attached to our artificial defensive structure, there is this solution to all our problems sitting there, available, where you live in support of this redeeming, reconciling and healing understanding of the human condition, which is the Transformed Way of Living. Well, if you have been able to follow the logic and imagine all that, you will, I suggest, have got your mind into a sufficiently honest position to actually realise and accept that this *is* the world's and your situation, and from there *actually* accept taking up the Transformed Way of Living—AND WHEN YOU DO, YOU WILL SUDDENLY EXPERIENCE HOW INCREDIBLY RELIEVING AND EXCITING THAT WAY OF LIVING IS, AND, AS I SAID, YOU WILL THEN NEVER WANT TO LIVE ANY OTHER WAY.

Cartoon by Michael Heath, *The Spectator*, 20 April 2024

[29] And, as I have said, the world *is* becoming so horribly messed up and dysfunctional that now, despite their historic practice of maintaining a positive, everything-is-okay outlook, more and more people are being able to let that delusion go and accept that this world-disintegrating situation (which Michael Heath's cartoon above so perfectly depicts) can't go on and that a fundamental change *has to occur*. Again, as Professor Prosen wrote, '**I think the fastest growing**

realization everywhere is that humanity can't go on the way it is going.' Just recently, the renowned podcaster Joe Rogan was talking about the world having reached this completely exhausted state where we desperately need a new, more loving and honest ideology that a human-condition-confronting-not-avoiding, truthful thinker like Jesus Christ would present, saying, 'I think as time rolls on people are going to understand the need to have some sort of belief in the sanctity of love and truth. And a lot of that [those values] comes from religion…Ethics are based on our moral compass and we all have one, but that's not necessarily true [anymore], we need Jesus!' (7 Feb. 2024). This led to another well-known podcaster, Russell Brand, agreeing with Rogan, saying, 'Yes, I do think we need a return of Jesus Christ…We need clarity, honesty, open-mindedness and a real alternative in the political sphere…I'd certainly like to see some new independent movements challenging these old institutions' (7 Mar. 2024).

[30] Two years earlier, in 2022, in one of his podcasts, the Scottish historian and archaeologist Neil Oliver expressed identical sentiments about how the political Right and Left have become meaningless for people and they are looking elsewhere, saying, 'So it's not Left and Right, it's become right and wrong, good and bad. For a lot of people…they're trying to understand what the rules [for a new ideology] might be, and where help might be found. So people are invoking myths like [the prophesised return of King] Arthur and people who had faith in [the prophesised 'second coming' of] Jesus Christ, both those stories take you to the same place that, "cometh the hour, cometh the man". People are looking to and hoping for a time when an Arthur or a Jesus will come back to right the wrong, it's amazing! Many people all around the world are feeling betrayed and let down by those [leaders and institutions] to whom previously they would've looked for safety. It's suddenly like finding a loving parent revealed as the opposite of a loving parent, an abusive parent, and people's faith in the structures and the institutions of society have been rocked and for a lot of people that faith might never come back, and they're looking to put their faith in something else. I think that's why you get a lot of people talking about alternatives, because a lot of people want to start again, a lot of people want to start alternative institutions' (3 Sep. 2022). The YouTuber and author Kevin A. MacLean similarly pleaded

for the return of King Arthur at the end of his documentary *Who Are The Welsh?*, saying, **'The need for Arthur's return couldn't be greater, and I'll be watching for him'** (28 Jan. 2023). I talk about how the finding of the human-race-saving redeeming explanation of our corrupted human condition represents the fulfilment of the prophecy of the return of King Arthur-like, relatively innocent, Celtic, soul-strong, denial-defiant courage and leadership in paragraph 1036 of my book *FREEDOM*, and about the fulfilment of the prophecy of the 'second coming' of Christ-like, innocent, sound and loving truthfulness in paragraph 1278 of *FREEDOM*, and in Freedom Essay 39.

[31] So, despite all the historic denial, more and more people are waking up from all the delusion and denial that has owned human life and are admitting this end play, human-condition-stricken situation *has* now arrived. And once a person understands this information, understands that the human condition has been explained, they actually know that there is another way we can live; they actually know that **'a real alternative' 'new independent movement'** based on **'the sanctity of love and truth'** that Rogan and Brand called for, and an **'alternative institution'**, **'not Left and Right'** but based on **'right and wrong, good and bad'** values that Oliver called for, *is* available, which is *exactly* what the Transformed Way of Living represents, and so when such a person then struggles to let go of their old artificial defensive structure, they need to listen to this talk that very clearly explains the Mexican Standoff and how to overcome it.

[32] Yes, the truth absolutely is that the human race has reached this end play situation where the old aggressive, dishonest, human-condition-denying, power and glory way of living is finished with, it's no longer viable, and we need an alternative, and, blow me down, there *is* an alternative that is all-solving and all-wonderful, so everyone *has* to let go of that old absolutely exhausted way of living and have Darling's great **'change of heart'** and take up the alternative Transformed Way of Living—and when they clearly face and accept that truth and as a result actually decide to, and do, take up that way of living, they will, at that moment, suddenly experience

how relieving and exciting it is to be living in the all-solving and all-exciting Transformed Way of Living!

[33] Just how exciting the Transformed Way of Living is can be gauged by the truth expressed in all the optimistic songs we've ever sung, and all the beautiful poetry we've ever written, and all the wonderful paintings we've ever painted—if you burrow down into what they are immersed in, what they are dreaming of, it is this time when we can let go of our old wretched way of living and become part of the warm and light, sun-drenched, human-condition-free, redeemed and reconciled new world. Even songs and literature and art that are not exploding with hope, optimism, anticipation and excitement (like, for example, 'rock and roll' music is, or the incredibly innocent, soul-alive vibrant paintings of the Australian Aboriginal artist <u>Emily Kame Kngwarreye</u> are, or the sensationally beautiful paintings of <u>Paul Gauguin</u> are) and instead are about suffering and hardship, the actual hope behind that agony is of one day finding understanding of the human condition that will end all of that agony. If you want to hear some fantastic anticipation-of-a-human-condition-resolved-new-world, exploding-with-optimism-and-excitement, 'let's-get-out-of-here', 'rock and roll' music, I don't know of any greater example (except maybe for the other early, before the now extreme angst of the human condition flooded the world, music by <u>Little Richard</u>, <u>Chuck Berry</u> and <u>Elvis Presley</u>—what did <u>John Lennon</u> say about Elvis, he said, **'Before Elvis there was nothing'**, everyone was asleep in an alienated trance!) than the live recording of <u>Jerry Lee Lewis</u>'s April 5, 1964 performance at the Star Club in Hamburg, Germany, especially the tracks *Long Tall Sally* and *Hound Dog*. It **'is regarded by many music journalists as one of the wildest and greatest rock and roll concert albums ever'** and that Jerry Lee **'sounds possessed'** and was **'rocking harder than anybody had before or since…words can't describe the music'** (Wikipedia; see www.wtmsources.com/125). My Freedom Essay 44: *Art makes the invisible visible* and F. Essay 45: *Prophetic songs* reveal the truth about this deeper meaning of optimism and hope in art and music, in fact in all of our creative expressions.

[34] There are just 'no dags on it', to use an Australian expression; there are no negatives in this new Transformed Way of Living. It is a totally effective solution to all of your and the world's problems, and so it *is* all-exciting and all-satisfying, and once your mind has got itself into a position where it is finally prepared to actually consider taking up, and then actually takes up, that way of living, your mind will at that moment experience how fabulously wonderful it is. I'm not kidding you—you will go from a state of likely being extremely distressed and depressed about the state of the world and of yourself, to a state of incredible optimism, excitement and happiness. That is the truth, that is how wonderful what is on offer is, and once you get just some experience of the relief and excitement, just a taste of what it's like to live in this state free of the agony of the human condition through living in support of this understanding that reconciles the human condition, <u>then you will finally understand how this Transformed Way of Living actually works, and then you can live off that experience all day, every day, forever!</u>

[35] But I'm just saying that you can't experience that freedom and excitement, you can't sample that, you can't taste that, whilst you're still locked onto your old artificial defence and reinforcement structure. You can't embrace the new while you're attached to the old; you can't, as they say, **'serve two masters'** (Bible, Matt. 6:24). Many people who understand the explanation for the human condition talk about the Transformed State, and they write in our Facebook Group about it, and they write and talk about the Mexican Standoff, and they write posts about how true and good this information is, and how wonderful it is, but I can tell—actually anybody can tell, 'reading between the lines' as it were—that they're not transformed yet, they're not full of the excitement yet. They're still stuck in no man's land between the two ways of living. So what I'm putting forward is a way to end that Mexican Standoff and discover how all-freeing and all-solving and all-exciting this new way of living is; the way to become really thrilled by it. It is the way to turn

your life around from seeing no real future for the world, where
every television program you look at is full of madness, suffering
and dishonesty, where every person's pushing their own 'bullshit
barrow', and the whole world is fake, and everywhere there is just
more and more devastation and destruction; stories of countries at
war and people in crisis, and refugees, and corruption and madness
everywhere you look, okay—and so that's a very depressing, in fact
terrifying (most especially for the sensitive souls of young people),
state of affairs. And your own world: if you understand the human
condition you understand how parents couldn't give their offspring
the reinforcement that their instincts expected because we have this
innocent soul, and so virtually everyone's lives are crippled as a result
of that experience. As this further depiction of the disintegration
of our world shows, it is very, very clear that we have reached the
absolute end point in the way we have been living!

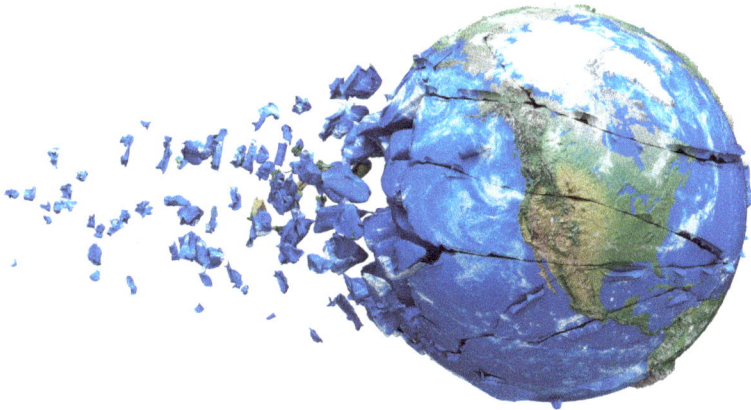

gurzart/Shutterstock

[36] So it is a desperate situation everywhere, and I'm trying to get
everyone to imagine extrapolating that situation to a state where
you can clearly, in a denial-and-delusion-free way, see that we have
reached the end point where that old way of living is no longer

sustainable for the human race, or for you individually, and that there *is* a solution that you and everyone can and now must take up. Because then, and only then I suggest, when you really and truly, in your heart of hearts, actually tolerate the idea, genuinely tolerate the idea, genuinely inhabit the idea of abandoning your old artificial power and glory way of living and take up living in a world free of the human condition where everyone is living self-lessly in support of this human-race-transforming understanding that solves everything, then you will have arrived in the heaven of **'the vision splendid of the sunlit plains extended'** (to quote Australia's greatest poet, <u>Banjo Paterson</u>) of a human-condition-free world for the human race!

© 2024 Fedmex Pty Ltd

[37] Honestly, it'll flip your situation from a state of constant ambivalence and struggle and desperation to a state of absolutely amazing excitement. That is not an exaggeration, that is the truth, but people can't see it, and the reason they can't see it easily is because they still have a foot in both camps; they're still actually determinedly holding on to their old power-and-glory-ego-castle way of sustaining themselves. So they can follow the logic of the

new Transformed Way of Living but they can't inhabit and thus experience the freedom, relief and excitement of it—and again, I'm suggesting the way to experience and know that relief and excitement is to follow the logic that makes it perfectly clear that we do have to have this change, and that we have got sitting there in front of us the perfect, flawless change, the perfect, flawless solution, to all of your and the world's problems. We don't have to all learn to meditate to fix the world, or support the utterly dishonest, human-condition-denying, no-truthful-biological-insight-into-any-thing, fake, dogma-not-understanding-based, 'regimented', pseudo idealistic, 'politically correct', 'woke', bullshit, makes-you-feel-good-but-takes-us-backwards-away-from-finding-self-understanding, culture, or any of the other false starts to a 'new age' that obviously can't and won't ever work (see my book *Death By Dogma*). There had to be an honest human-condition-confronting-and-solving, first principle-based, biological explanation for why we humans are good and not bad—a psychologically relieving understanding of ourselves, of our corrupted condition—for our minds to ACTUALLY be able to be free of the agony of the human condition.

[38] So we can all live in support of this all-relieving and all-exciting human-condition-solved REAL solution to the world's problems. And again, if you're talking to someone who understands this information you can actually know where they are on the spectrum between being infected with the excitement of what it is to live free of the human condition, and to still be living firmly attached to their old artificial defence structure.

[39] So I'm just suggesting that if you extrapolate where you are at, and where the world is at, you will get to the point where you will accept that our existing artificial, power and glory, 'I'm a legend'-reinforcement system is finished with, and then and only then, I'm suggesting, will you genuinely consider and then actually take up support of the Transformed Way of Living and by so doing experience how fabulously relieving and exciting it is living that way. It's just sitting there available, this alternative way of living

where we live in support of the actual understanding of the human condition, and it is 100% bullet-proof, 100% true, 100% exciting, and 100% will solve all of our problems—so, really and truly, how incredibly exciting is it to be able to live that way!

[40]While we are still living in this early pioneer stage where the Transformed Way of Living hasn't yet caught on in a big way, most people who recognise the truth of these understandings of the human condition get stuck halfway in the Mexican Standoff between the two poles of maintaining their existing resigned, denial-complying, competitive, power and glory way of validating and feeling good about themselves, and adopting the new all-relieving and all-exciting Transformed Way of Living. But I'm telling you, this infected-with-excitement, new Transformed Life is going to catch on everywhere! As I have mentioned, people get excited with 'rock and roll' music because what is driving the excitement in the music is the anticipation of a time when we would be free of the human condition, but honestly 'rock and roll' music only touches the sides of how exciting this human-condition-relieved new world is going to be, and you can work out from everything I have said that that is true! All great art, great literature, great music, great creativity—and also, as I explain in paragraphs 757 & 786 of *FREEDOM*, all of our 'fall-in-love' romance, which has been a huge part of our lives—is about dreaming of, and living for, and fighting for, and tapping into this potential, this inchoate hope, faith and trust that we humans have held deep within ourselves of one day being free of the soul-corrupted, angry, egocentric and alienated human condition—and that great anticipation, that fabulous, absolutely incredible dream, has actually come true and arrived! The excitement and the beauty in great paintings, theatre, ballet, literature, or whatever form of expression, is that they're all expressing the dream of this time that has actually arrived, and it *is* absolutely and truly 100% exciting and beautiful! There is absolutely no need any longer to stay living in the old tortured, pressured, brutal, hateful, dishonest, crippled, immensely-heroic-but-wretched world! I don't

live in that world, I just live in this other soul-full, cooperative, self-less and loving, human-condition-free world all the time because in my mind, ever since I was a child, I have known the difference between the two; between living in the artificial, superficial, trashy, dishonest, ugly, power-and-glory-obsessed way of living and this other authentic world. This other potential real world is so infinitely exciting and fabulous that the other world has no attraction for me at all! It's invisible to my eyes—apart from all the suffering in it—it's of no significance to me. There is this other fabulous place of potential that I live in and live for—the return of our species' original cooperative, selfless and loving existence, as dreamed of and prayed for in *The Lord's Prayer*: **'Your** [Godly, integrated, peaceful] **kingdom come, your** [integrative] **will be done on earth as it is in heaven** [and once existed on Earth]' (Bible, Matt. 6:10 & Luke 11:2). So that other place is where I get my energy from, that's what I live for, it's just 100% satisfying, 100% exciting, and this way of living is going to catch on because we have rid ourselves of the guilt we have felt for corrupting our soul that was stopping us from living there. So that all-exciting and all-satisfying way of living is now available to *everybody* (see my Biography, and more particularly my book *How Laurens van der Post Saved The World*, for an in-depth description of the exceptionally nurtured and exceptionally soul-sheltered origins of the orientation of my life). And I'm just suggesting, if you just do some extrapolation, that your situation and the human race's situation is going to lead to this end play state where we *have to* make a choice, then you're forced to actually consider and then actually try letting go of that finished-with way of living and instead live for this all-redeeming and all-healing and all-satisfying understanding of the human condition; and then, and only then, will you actually be able to discover the excitement of this human-condition-free new world for the human race!

[41] In Christianity, for example, if you live in support of Christ you're living in support of the embodiment of the ideals, and, as St Paul realised and taught, by doing that, by living in support of

Christ rather than in support of your corrupted condition, you are able to free yourself from your corrupted condition. But, as I explain in paragraphs 42-50 of *Death by Dogma*, Christianity and the other great religions were only stepping stones, albeit critically important ones, towards this ultimate understanding-of-the-human-condition-based free life. And yet, as I explain in paragraphs 1198-1200 of *FREEDOM*, in the case of Christianity, the relief St Paul derived from deciding to live in support of Christ was so immensely exciting he fell off his donkey and, metaphorically speaking, went blind for three days.

St Paul falling off his donkey and going blind with ecstatic relief
after he let go of his struggle with the human condition

[42] If we look more closely at what happened, St Paul, or Saul as he was known then, was an extremely psychologically upset, angry, egocentric and alienated person who was on his way to Damascus where he planned to persuade the authorities to destroy the apostles, such was his fury towards the truth about our corrupted human condition that Christ, through his sound words and life, had dared to reveal, and which this fledgling group of Christians were deeply appreciating and loving. However, while riding along on his donkey

in a seething, **'murderous'** (Bible, Acts 9:1) rage about Christ and his followers, Saul had an epiphany, the effect of which was so incredible that it was to change the world and basically save the human race from unbearable levels of upset. Saul's anger was obviously very great, but so was the level of self-hatred he would have felt deep within himself for being such a brutally angry person. This extreme self-hatred led Saul to think, 'What if I just live in support of this extraordinary human [Jesus Christ] that Stephen was prepared to give his life for, that Stephen wouldn't disown?' (see Bible, Acts 7:54-59), and at that moment Saul *actually* considered what doing that would be like, and what happened when he *actually* did that was that immense relief and excitement flooded through him. As it says in the Bible, he was **'born again'** (John 3:3) from his corrupted condition; he had **'crossed over from death to life'** (John 5:24); he had become **'a new creation'** where **'the old has gone, the new has come!'** (2 Cor. 5:17).

[43] THE VERY IMPORTANT POINT here, that everyone needs to understand because it illustrates everything I have been saying, is that Saul didn't decide to live in support of Christ because of how extremely relieving and exciting that would be; no, he was *driven* to living in support of Christ because he had enough of how angry and destructive his extremely upset self was. It was only *after* he was *driven* to supporting Christ rather than supporting his upset self that he discovered how 'fall-off-his-donkey-and-go-blind-with-relief'-wonderful it was abandoning his upset way of living. He had to be *pushed* into making that change before he discovered the fabulous transforming effect of doing that—he was not *pulled* into making that change by the wonderful relief that doing that would be. AND THIS IS EXACTLY THE SAME SITUATION PEOPLE IN THE MEXICAN STANDOFF ARE NOW IN. They are so attached to all their defensive denials and all their power and glory self-validations that they are unable to know how fabulous it is to let that go. They have to be *pushed* there by getting the logic of their and the world's situation clearly lined up in their mind before they will let go of their artificial

defensive structure, and then, and only then, will they—like St
Paul—discover the immense relief and excitement of doing that.
Their determined attachment to their denials and delusions and power-
and-glory-ego-castle blocks them from discovering the wonders of
the Transformed Way of Living! They should be able to be *pulled*
to adopting the astronomically relieving and exciting Transformed
Way of Living, but their attachment to their old defensive structure
means they have to be *pushed* there through thinking very cleanly
about their situation. In St Paul's equivalent situation, he wasn't
expecting to fall off his donkey with relief, he only discovered that
wonderful effect when he was *pushed* into letting go of his upset
way of living. It is a ridiculous situation that those in the Mexican
Standoff are in because the Transformed Way of Living is so completely, utterly all-satisfying and all-wonderful, but such is people's
attachment to their, in truth, feeble artificial reinforcement structure
that they can't access it! They are 'caught' in the Catch-22 situation
described earlier (in par. 25)!

[44] I should mention that Christ's 12 disciples also experienced
the problem of the Mexican Standoff where they initially strug-
gled to overcome their attachment to their existing self-focused,
self-validating, egocentric way of reinforcing themselves and see
the fabulous benefit for their overly upset condition of letting that
way of reinforcing themselves go and instead deferring to and living
through their support of Christ (for the explanation of Christ and
Christianity, see Freedom Essay 39). Just as St Paul initially failed
to be 'pulled' by the fabulous benefit of living through Christ, so
too did the disciples initially fail to be 'pulled' by that benefit. They
venerated and worshipped Christ, but that was not the point; as
mentioned, the real point of Christianity is to let your overly-upset,
extremely embattled, artificially-reinforcing, self-preoccupied
defensive structure go and instead live gloriously free of all that
ugly, wretched preoccupation by deferring to and supporting the
sound words and life of Christ, who acknowledged and taught that
'I am the light of the world [I am a soul-alive, uncorrupted person]. **Whoever**

follows me will never walk in darkness [never have to live in a soul-destroyed and truth-denying, corrupted, angry, egocentric and alienated condition], **but will have the light of life'** (John 8:12). But the disciples initially resisted abandoning their attachment to the **'darkness'** of their corrupted, angry, egocentric and alienated way of living, actually complaining that **'This is a hard teaching. Who can accept it?'** (John 6:60)! <u>So, just as people initially resist letting go of their once necessary but now unnecessary agonising, ugly and wretched artificial reinforcement structure and taking up the glorious Transformed Way of Living where you live a fabulous existence in support of the actual redeeming and relieving, first-principle-based, biological understanding of ourselves, St Paul and the disciples also initially failed to see and take up the wonderful opportunity of letting go of their angry, egocentric and alienated structure and instead putting their faith in Christ. Of course, the Transformed Way of Living is infinitely superior to putting your *faith* in an exceptionally sound person, because the Transformed Way of Living is based on living in support of the actual redeeming, biological *understanding* of our corrupted condition. As I have often said, what is on offer now is the end of dogma, faith and belief and the beginning of knowing!</u>

[45] Yes, as I said earlier, Christianity and the other great religions were critically important 'stepping stones' on humanity's journey towards this ultimate finding of the explanation of the human condition that makes the actual rehabilitation of the human race possible. Religions provided solace, comfort, a way for the upset human race to live in relative freedom from their corrupted condition and by so doing bought time for the actual understanding of the human condition to be found. In the case of Christianity, Christ recognised this purpose. Again, in his *Lord's Prayer* Christ instructed us to pray for the time when **'Your** [Godly, integrated, peaceful] **kingdom come, your** [integrative] **will be done on earth as it is in heaven** [and once existed on Earth].' He also looked forward to the time when **'another Counsellor to be with you forever—the Spirit of truth** [the denial-free, truthful, first-principle-based, scientific understanding]**...will teach you all**

things and will remind you of everything [all the denial-free truths] **I have said to you'** (John 14:16, 17, 26). He similarly said he looked forward to when, instead of being restricted to **'speaking figuratively'**, we **'will no longer use this kind of language but will** [be able to] **tell you plainly about my Father** [be able to explain the world of Integrative Meaning in denial-free, human-condition-reconciled, compassionate, understandable, rational, first principle, scientific terms. See Freedom Essay 23 for the explanation of the Negative Entropy driven process of Integrative Meaning that we have personified as 'God']' (John 16:25). So religions like Christianity were only a step, albeit a major one, along the way to finding this ultimate solution to our corrupted condition, and yet even that was all-exciting, leading Saul to 'fall off his donkey', so imagine how much more exciting it is now that we have the understanding that will actually FREE the human race from the human condition!

[46] So, I'm saying, if St Paul fell off his donkey with the excitement of being able to let go of the upset state of the human condition in that unfinished, human-condition-was-still-to-be-explained stage, then now that we have the finished state, now that we have found the actual redeeming understanding of the human condition that we always intuitively hoped for and can now live in support of, how *infinitely* more exciting is that! St Paul sold Christianity by comparing it with living under the rule of Moses's Ten Commandments, because he said if the latter was considered glorious how much more glorious is it to be on the side of righteousness rather than living through the discipline of living in fear of punishment for disobeying the Ten Commandments—that 'if you do this you will be punished, if you do that you'll be punished'. As I wrote in paragraph 1215 of *FREEDOM*, St Paul gave what was possibly the best sales pitch for born-again, Prophet-Focused Religious life when he wrote, **'Now if the ministry that brought death, which was engraved in letters on stone** [Moses's Ten Commandments that were enforced by the threat of punishment], **came with glory** [because they brought society back from the brink of destruction]...**fading though it was** [there was no sustaining positive in having discipline imposed on you], **will not the ministry of the Spirit be**

even more glorious? If the ministry that condemns men is glorious, how much more glorious is the ministry that brings righteousness [that makes you feel good about yourself]! **For what was glorious has no glory now in comparison with the surpassing glory'** (Bible, 2 Cor. 3:7-10). So St Paul sold that way of living by comparing Christianity to the Ten Commandments as a system that offered not punishment but relief for humans' corrupted condition. Well, given this is the *actual understanding*, it's so, so much more exciting than that—if that was exciting then what's on offer now *is* infinitely more exciting, this is the **'surpassing glory'** of all **'surpassing glories'**!!

[47] So that is the new, true situation. Nobody has to live back in that old finished-with, exhausted, spent, embattled, brutal world. Everybody can, every moment of the day, just live absolutely flooded with their mind thinking about this wonderful new world that they're living in support of and the excitement it brings. And that changes everything. It changes your life from having a bet both ways; living in the Mexican Standoff with one foot in one camp and one in the other, where you are saying, 'Yeah the information is true and I should let my artificial ways of validating myself go, but I have to say I'm pretty attached to my denials and castle of wins'!

[48] The 'How To Help' and 'The Transformation' sections on our website begin with my description of how, if I was destitute and living under a bridge and someone gave me this understanding of the human condition, I would get the basic necessities of my life in place of somewhere to live and a job that would earn me enough to live, and then hand out leaflets and support our online advertising; and I said that by doing those two explanation-of-the-human-condition-promoting things I would feel that my life was completely meaningful, fulfilled and happy. I'm aware that often people read that and think 'That's a bit harsh, a bit strident, even fanatical, it's not for me; it's so unrealistic, not really meaningful and fulfilling because I have a family to look after, children to school, and I need a decent house and environment to live in, so I need a high income

earning job to be able to do and have these things.' However, as I have emphasised many times, such as in paragraph 20 of *The Shock of Change*, 'the focus of the Transformed Way of Living where you leave the old world of artificial reinforcement isn't on giving up your possessions or walking the streets in sackcloth in self-denial and servitude. We're talking about a change of mindset that can have an effect on your priorities, which can affect your choice of possessions and so forth, but the focus isn't on self-deprivation.' Yes, we still have children and homes and go to work but *how* we go about those activities changes dramatically. The orientation, focus and preoccupation of our lives changes from attacking any perceived criticism of ourselves, denying the corrupted state of the human condition, and gaining as much power, fame, fortune and glory artificial reinforcement of ourselves as we can. We leave our angry, alienated and egocentric competitive selfish and aggressive way of living go and live cooperatively, selflessly and lovingly in support of the redeeming, reconciling and rehabilitating understanding of the human condition. The whole *focus* of our lives changes.

[49]The real reason people find my transformation-from-living-under-a-bridge story strident and unrealistic is actually because they are still in the Mexican Standoff and not seeing the magnificent opportunity we now have to save the human race from unthinkable, absolutely horrific amounts of human suffering from terminal levels of psychosis, and then our species' extinction! If your house is burning down (and our world *is* disintegrating) and you're given a hose that will put the fire out (given the redeeming understanding of the human condition) you go and put the fire out; isn't that what you do? When Stephen wouldn't disown Christ (which I talked about earlier in par. 42) people couldn't relate to someone being that committed, but that is only because their Mexican Standoff attachment to their self-obsessed existence meant they were not seeing the situation clearly like Stephen was where living in support of Christ was the way to relieve the human race of the unbearable horror of the human condition—which, in Stephen's case, his commitment actually *directly* had the effect of

doing because it inspired St Paul's human-race-saving conversion (which, again, is what I pointed out in par. 42)!

[50]I wasn't being 'a bit strident' in my transformation-from-living-under-a-bridge story, I was just living in a state of ecstatic excitement and relief that I was able to focus, not on being party to all manner of dishonest denial and to making myself feel that I am some sort of a legend, but on fixing up all the horror and suffering of the human condition forever because at last we *really* can! In the short videos at the end of this presentation of my 'Sermon On The Beach', the ecstasy and focus of Michael, Ales and Nikola is not a response to one of those deluded, pseudo idealistic, dogmatic, no-understanding-based, 'regimented', 'dislik[able]' (as Sir James Darling described them) New Age/Politically Correct/Woke *false starts* that we have all had enough of because they are only leading us deeper into denial and darkness, but clear-minded vision of what understanding of the human condition now *actually* makes possible! Yes, all the pseudo idealistic false starts to a human-condition-free world have horribly discredited the real start to a human-condition-free world that is now genuinely possible. Now that we have the redeeming understanding of our corrupted condition, letting go of our attachment to our self-focused, egocentric way of living becomes not only possible but a necessity, whereas letting go of that attachment when the upsetting, heroic battle to find that understanding still had to be won, that all the pseudo idealistic false starts to a human-condition-free world practised, was a complete *fraud*. It is true that letting go of our self-focused, egocentric way of living and instead living a non-self-focused, non-egocentric transformed life has been what members of religions, and even cults, have done—in their case, by deferring to a deity or some form of mind control—but that is where the superficial similarity ends. The new Transformed Way of Living where we live in support of the actual understanding of the human condition is the first *real, genuine, legitimate, imperative* transformation from living a self-focused, egocentric life. Yes, most importantly, as I explain in

some detail in my book *Death By Dogma*, while all previous forms of abandoning our upset angry, egocentric and alienated life were fundamentally irresponsible and pseudo idealistic, abandoning our upset, egocentrically embattled, denial practising life is now not only legitimate, it is the *only* way to live.

[51] The overall truth is that the discreditation of the Transformed Way of Living from all the PSEUDO IDEALISTIC FALSE STARTS to living a human-condition-free life, the DEAF EFFECT difficulty of accessing the redeeming explanation of the human condition, and the absurd MEXICAN STANDOFF resistance to taking up the Transformed Way of Living are what have been stopping the instant liberation of the whole human race from the human condition that understanding of the human condition now makes possible! The fact *really* is that the human race is now free, everyone can just walk out of the old embattled, human-condition-stricken world forever! It's so simple, but because of the three blockages just mentioned, people are struggling to wake up to that reality. We have been living in so much darkness for so long it does seem absurd to say it, but it is actually true—anyone and everyone can, right now, just walk out of and leave the old desperately dark, embattled, human-condition-stricken world forever!

[52] And with regard to the Mexican Standoff, which is what I've been focusing on in this presentation, it can now be seen as a horrible affliction blocking people's ability to see the utter magnificence of what's on offer for the human race! What did Sir James Darling say in that quote of his that I included earlier in paragraph 27—he said 'selfishness' is 'ultimately' a 'destructive force in a society' and so even though 'the change of heart' to living selflessly is something 'we refuse to make', having that 'cure' 'for selfishness' of 'the change of heart' to living selflessly 'is the choice, believe me, for each one of us, and we have not much time in which to make it. The need for decision [to have a 'change of heart' and live selflessly] is serious and urgent, and the sands [of time] are running out'—and those sands *have*

run out much, much, much more in the 74 years since Sir James gave that speech!! As I explained earlier, we have had to 'refuse to make' 'the change of heart' to living selflessly while we still had to persevere with our corrupting search for understanding of our divisive, corrupted, 'fallen' human condition, but now that we have found that understanding the human race can have that 'cure' 'for selfishness' of 'the change of heart' to living selflessly. I didn't mention it when I included Sir James's quote earlier, but Sir James actually knew that the ability to have the 'change of heart' to living selflessly depended on solving the human condition, which, astonishingly, he actually dedicated his whole education system at Geelong Grammar School to achieving, saying in speeches that (the under-linings are my emphasis) 'Only in the light of…a discovered purpose can we lead Australia into an attitude of mind which is prepared for sacrifice and service…in seeking for such purpose it will be necessary to seek below the surface…[for the] thoughts which do lie too deep for tears [which are those thoughts that for most people are so depressing and confronting they can't go near them, namely thoughts about our corrupted human condition]… Only so can we come to a better understanding of life, to answer even the all-important question: "What is man that thou art mindful of him [why is human behaviour so often less than ideal], and the son of man that thou visitest him?" [when Christ's behaviour by contrast was sound and ideal— Sir James has here clearly stated that the 'all-important question' that we have 'to answer' is the issue of our species' less-than-ideal, soul-corrupted, 'fallen' human condition.] For to exclude that question from the study of evolution…[is] surely as futile as to talk theology and to forget evolution [a biologist is going to be needed]? There must be a complete answer; there must be coherence and sense in the universe; and, until we find it, our think-ing is degenerated into disintegration, and our existence fragmented into a rubbish-heap of shreds and patches, with coherence, significance, and growth impossible, our compass-bearings lost, and civilization foundering…Canon Raven, says, the future lies not with the predatory [selfish] and the immune [alienated] but with the sensitive [innocent/sound] who live dangerously [defy the vast, Earth-cloaking-and-Earth-choking, immensely insecure, deathly-dark world of denial]. **It should be the prime object of education... to develop this sensitivity...the truly sensitive mind is both susceptible and penetrating: it is open to new ideas, and it seeks truth at the bottom of the well. It is the development of this sort of mind which it should be the object of the educational process to cultivate'** — **'each of us should regard our lives as pledged to the one paramount purpose of saving the world...the sands of time are running out'**. So at Geelong Grammar School Sir James actually set about finding and preserving the innocence needed to achieve the goal of solving the human condition—and, astonishment on top of astonishment, he actually succeeded because that is what his absolutely incredible vision achieved with the finding of the explanation of the human condition that I am presenting! (You can access the source of these quotes, and read more about Sir James Darling's superlative-beyond-all-superlatives vision to create a Plato's **'philosopher kings'**-inspired school to actually solve the human condition, in Part 5 of *The Great Transformation*.) So, we have the explanation of the human condition now that makes the **'change of heart'** to living selflessly possible, so everyone can, and now must because, as Sir James said, the **'decision is serious and urgent'**!

[53] Yes, what is on offer now is actually, truly, the *real* transformation of the human race that we've always dreamed of. <u>This *is* the most exciting event in human history—and it's only a matter of time before everyone wakes up and realises that</u>. And that great awakening is starting to happen. If you look at what's occurring on the WTM's Facebook Group, it's March 2024 and already you can see people becoming incredibly excited, handing out leaflets and putting our 'Fix The World' stickers on their cars, etc, etc, and that enthusiasm is unstoppable. (All WTM products are freely available at www.humancondition.com/products.) This is the time the great denial-free thinking prophet, scientist and philosopher, <u>Teilhard de Chardin</u>, anticipated when he wrote, **'The Truth has to only appear once...for it to be impossible for anything ever to prevent it from spreading universally and setting everything ablaze'** (*Let Me Explain*, 1966; tr. René Hague et al., 1970, p.159 of 189)!

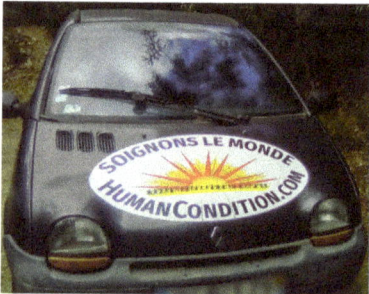

'Fix The World' in French

FIX THE WORLD
HUMANCONDITION.COM

THE TRUTH IS the levels of dysfunction and devastation everywhere now are so great we simply have to find the redeeming and psychologically healing understanding of our deeply troubled human condition — or all is lost!

Well, astonishing as it is, that 11th hour breakthrough has been achieved! As Professor Harry Prosen, a former president of the Canadian Psychiatric Association, has said, "I have no doubt Australian biologist Jeremy Griffith's explanation of the human condition in his book *FREEDOM* is the holy grail of insight we have sought for the psychological rehabilitation of the human race."

Just watch the now famous interview with Griffith at **HumanCondition.com** (or scan the QR code) and you will know it's true!

(Please turn over to read many more commendations from world-leading scientists for Griffith's treatise.)

HumanCondition.com

Tim Macartney-Snape on the summit of Mt. Everest

THIS IS THE BEGINNING OF THE WORLD WE ALWAYS HOPED FOR

World Transformation Movement
www.HumanCondition.com

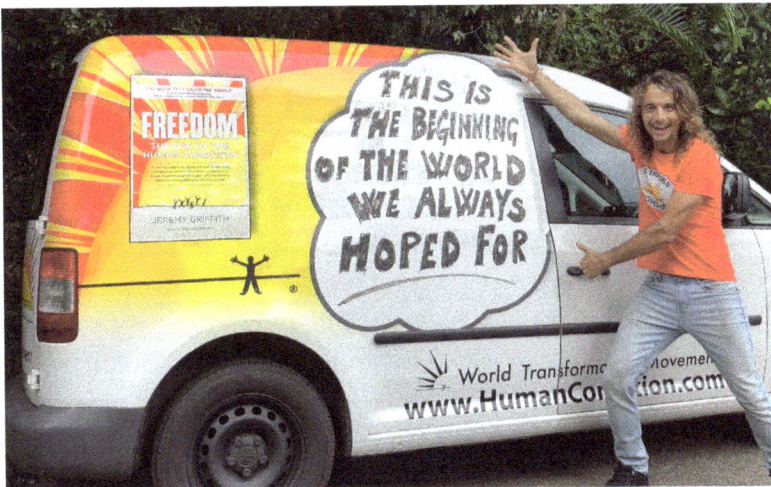

⁵⁴But it is true, if you look at everybody involved with this information you can feel how much they've actually cottoned on to how this actually works. There are people like <u>Tony Gowing</u>, and many others, who are fully immersed in this new way of living, and fully flooded with the excitement, energy and functionality that comes from that, and others still variously in the Mexican Standoff but on their way to fully living the Transformed Life. So it's on, the great break-out of humanity from the agony of the human condition—humanity IS coming home to peace and happiness at last!

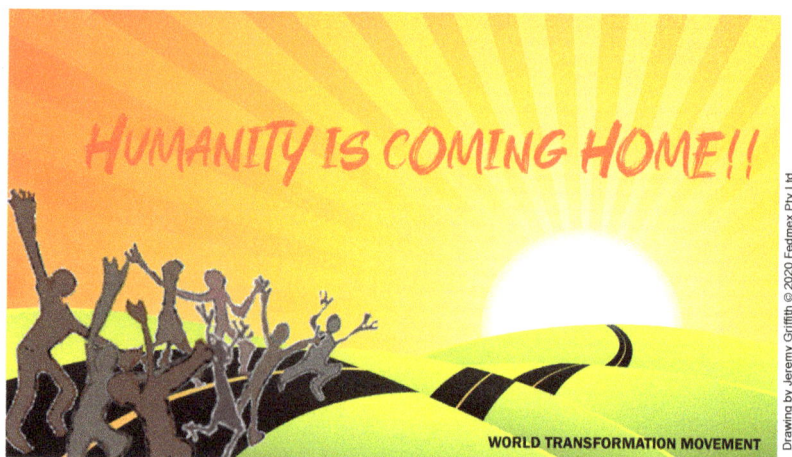

HUMANITY IS COMING HOME!!

WORLD TRANSFORMATION MOVEMENT

Drawing by Jeremy Griffith © 2020 Fedmex Pty Ltd

⁵⁵In our alienated state we don't realise how obvious it is that the world is completely false and that one day that falseness has to be exposed, but to a non-alienated, denial-free mind those are obvious truths. Alienated, human-condition-denying minds don't acknowledge that humans suffer from the corrupted state of the human condition, so for them there is nothing to be liberated from. Denial-free thinking prophets, on the other hand, recognise our corrupted condition and look forward to the day of liberation from that horrible false existence. They look forward to the time when

understanding of the human condition would be found and the false
world virtually everyone's living in will be exposed, and when that
happens how <u>people will abandon that false, angry, egocentric and
alienated way of living in droves</u>. Despite it being thousands of years
in the future for him, the great denial-free-thinking, clear-sighted
Biblical prophet <u>Joel</u> dreamed of, and lived for—and gave this
perfect description of—this time when the redeeming understanding
of the human condition would be found and the human race would
be liberated and an unstoppable host of transformed humans would
appear to sweep away all the madness and suffering from planet
Earth: **'Like dawn spreading across the mountains a large and mighty army
comes, such as never was of old nor ever will be in ages to come…Before them
the land is like the garden of Eden, behind them, a desert waste—nothing escapes
them. They have the appearance of horses; they gallop along like cavalry. With
a noise like that of chariots…like a mighty army drawn up for battle. At the
sight of them, nations are in anguish; every face turns pale.** [The reconciling
truth is shockingly confronting and exposing at first, but those who have
progressed past the shock stage are overwhelmed with excitement about
being transformed and having the capacity to end human suffering and the
devastation happening on Earth.] **They charge like warriors; they scale walls
like soldiers. They all march in line, not swerving from their course. They do
not jostle each other** [Joel 2] …[and] **In that day the mountains will drip new
wine, and the hills will flow with milk; all the ravines…will run with water…
Their bloodguilt, which I** [Integrative Meaning/'God'] **have not pardoned, I
will pardon** [Joel 3].' Yes, as I have said in every main publication I have
ever written, going right back to my first book *Free* in 1988, **'soon
from one end of the horizon to the other will appear an army in its millions
to do battle with human suffering and its weapon will be understanding'.** <u>It
won't be long before everyone you walk past will be smiling and
happy because they know about the Transformed Way of Living
and how it saves the world. A great weight has been lifted off
humanity—its 2-million-year long guilt for having corrupted our</u>

all-loving soul—so there is just going to be immense relief and joy everywhere! Imagine that, relief and happiness everywhere! At last, everyone can finally breathe freely, their faces radiant with immense relief and joy! Yes, this eleventh hour breakthrough redeeming understanding of ourselves means the outlook of the human race is going to change from terror about the state of the world to sublime relief and happiness! And yes, truly as Tim Macartney-Snape, our deeply loved founding member and Patron of the WTM says, 'The WTM is an island of sanity in a sea of madness'! It won't be too long before the whole 2-million-year horror story of the human condition will be a distant memory for the human race; but, yes, in the meantime, every human, the whole human race, has a safe place to live; has an all-loving and all-caring and all-inspiring sanctuary in the WTM! To re-use in a whole new, even grander way, Christ's words from his sublime, central-to-all-his-teaching, human-race-saving 'Sermon On The Mount': the WTM is 'the light of the world...it gives light to everyone', allows everyone to 'be comforted' and part of 'the kingdom of heaven' (Matt. 5:4-14) of an *actual* human-condition-free world.

[56]So in finishing, I'm just suggesting if everyone imagines that the situation of the human race is going to get serious very quickly, and in fact that this end-play state *is* actually upon us, and that everyone has to make this choice between the old and the new, that there is no longer any possibility of sitting in the middle—then, as a result of that clear logic, you *actually* decide to take up the Transformed Way of Living, then, at that moment, you *will* discover how incredibly exciting it's going to be. And when you do, when you finally let that excitement in and experience it, then the old steel casing of living off artificial wins that you have been locked in will suddenly shatter and the light of immense truth, and immense relief, and immense happiness will flood into you,

and every exciting possibility you have ever dreamed of will open up for you, and for the whole human race! This is going to be the biggest, most exciting movement the world has ever seen! It is just going to sweep everything before it, be **'Like dawn spreading across the mountains'**, **'setting everything ablaze'** as Joel and de Chardin said. No one's going to be looking back, everyone's going to be looking forward to a magic new world for the human race! So LET'S GO, let's all now leave this hell hole forever!

[57]**Tony Gowing**: That completes Jeremy's fabulously inspirational 'Sermon On The Beach'!

[58] As I mentioned initially, I want to strongly recommend reading the transcript of this talk because in it Jeremy has added many extremely helpful and enlightening and important elaborations to what he has just said. So please don't miss out on reading the transcript because it's an absolute feast!

[59] Now I want to conclude this presentation with these short affirmations of everything Jeremy has said by <u>Michael Perry</u> from Alberta, Canada; <u>Ales Flisar</u> from Queensland, Australia; and <u>Nikola Tsivoglou</u> from Victoria, Australia. And, as you should fully understand by now, the excitement of Ales and the relief of Nikola is completely legitimate: they are not more of the sickeningly deluded, dogmatic, no-understanding-based, pseudo idealistic, New Age/PC/ Woke *false starts* to a human-condition-free life that we've all had enough of because they are only leading us deeper into denial and darkness and so **'all profess to dislike'**, as Sir James Darling said, rather they are responses to the **'complete answer'** that Sir James referred to of being able to *actually* understand the **'all-important question'**, that Sir James also referred to, of the human condition—which Michael's deeply felt presentation expresses the real magnitude and utter magnificence of. These are *genuinely* liberated responses to the *actual* biological understanding of ourselves that the human race has been tirelessly working its way towards finding.

Michael Perry's Affirmation

Michael Perry, Alberta, Canada, April 2024

[60] **Michael Perry**: [When Michael posted this video in the WTM Facebook Group, he introduced it with this text: "I am overcome with joy and appreciation for Jeremy's biological understanding of the human condition and this beyond wonderful World Transformation Movement which supports that life-changing, literally world saving understanding. Felt like stretching out in the sunshine and sharing the love/truth."]

[61] I just wanted to hop on here real quick because I'm having one of those moments—and I'm positive many, many of you can relate to what I'm feeling right now. Which is just, sometimes you're able to access the full gravity and magnitude of what's going on and it's

just completely absolute, and solid, and all-encompassing. And you can see the significance of what we're doing and what Jeremy has found. And you can see the whole—you know, so many people talk about the macro level—and you can see it. You can see everything right from innocence, right to our 'Garden of Eden'-like original instinctive past up until now. And you can see the full circle of it and how we had to suffer and find this understanding. And that feeling is, it's almost indescribable, and I know there's many of you who know what I'm talking about, but it's—the closest word if I can, like, download it into the language that we can currently speak—is 'heaven'. It's like heaven, God-knowing heaven on Earth.

[62] It's unbelievable, just to see the other side of it, and then you get, like, double excited because you know there's others out there who are feeling this and it's only gonna grow and grow and grow and grow and grow until every last human being on the planet knows, they know the truth. Man [shaking his head], anyway, short and sweet. I'm on a lunch break right now and I'm just enjoying nature. We get to enjoy nature now, we get to love nature and feel how it provides for us; we no longer feel condemned by nature, we can become friends with nature again and it doesn't feel silly to say that. You don't feel deluded, you feel secure in saying that we can become friends with nature.

[63] It's unbelievable, I love this movement so much, and I would be just a chaotic ball of answerless anxiety if it wasn't for this movement. I love this movement and I love all of you, and just what this represents. Alright, yeah, thanks for tuning in.

Ales Flisar's Affirmation

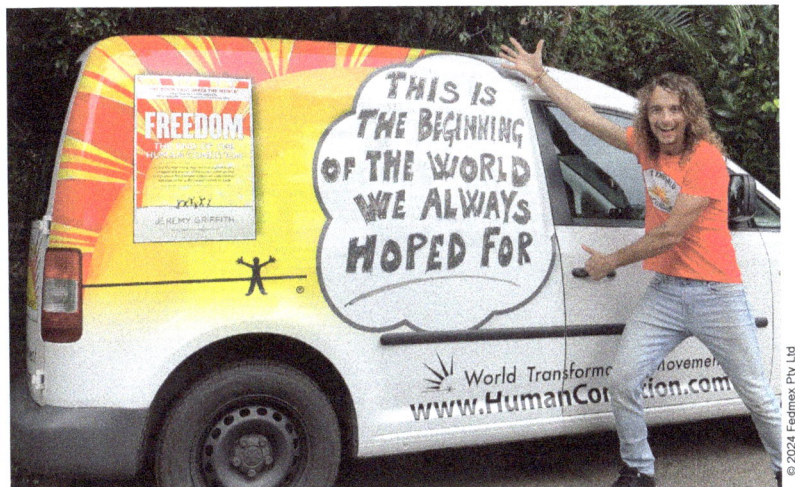

Ales Flisar with his van, Sunshine Coast, Queensland, Australia, May 2024

[64] **Ales Flisar** (from the Sunshine Coast (Australia) WTM Centre): I gotta say that the 'Beach' Video, I don't know, it just changed me.

[65] I'm just bloody excited about this. This is it, like you know. I'm talking to people about this. I'm bloody wearing this shirt now. I threw all my other shirts away! It's both sides; it's back and front—I'm a walking billboard now!

[66] I was camping here last night with a bunch of people and eventually started talking about some real stuff and everyone agreed that the world is in chaos. And I'm like, I'm not gonna focus on that. I'm gonna focus on the good stuff. I couldn't hide the excitement because it's here. It's within here [gesturing to his heart]. It's here. We can grab it. We can grasp the whole thing and go forward. This new world is good. I'm controlling myself here!

[67] Thanks Jeremy. All of you guys. I know you guys are shoulder to shoulder with me. Even though we are so far apart. I just had this vision: if we all came together at some point, we will—there's gonna be many, many, many more—we'll actually walk shoulder to shoulder, through the streets, through the towns, villages, wherever. [Note, this is identical to what the prophet Joel predicted would happen when understanding of the human condition was found.] That's what's happening right now, in my head I know you guys are there shoulder to shoulder. I knew that everyone's bloody excited about this [*Sermon On The Beach*] Video. I freaking felt all of you because we are on the same page, the same depth. You know, we're juggling between two worlds [in the 'Mexican Standoff']; but it can stop right now, right here. And yeah, so thanks Jeremy, and all of you. Bloody hell, this is good.

[68] **Genevieve Salter** (WTM founding member from the Sydney WTM Centre): How did you go, Ales, when you were talking last night? Did you get much feedback?

[69] **Ales**: No, everyone wanted to go to sleep because of the Deaf Effect. Some will engage because there's a lot of truth in that conversation. And I'm giving them examples from my life and it's so easy because I'm defended. I know I'm defended. And in the end, I gotta say it's all over, I don't need to prove my worth. I'm not chasing power, fame, fortune and glory no more. That's the good thing here—yeah, it was really good. But some people just changed the subject at some point and I knew when to stop. And that was it, you know, seeds planted for sure.

[70] I've got the shirt on every day. And my van, it's in everyone's face. It's right there; it's in their face. I open my phone, there's a QR code—let's go, straight into *THE Interview* because that's where I started.

[71] I'm out in nature, look, innocence all around me. I can finally see all this [nature]. And then we got this thing here [on my van]. "This is the beginning of the world we always hoped for."

[72] **Genevieve**: Awesome!

[73] **Ales**: And that's what I'm driving around in. It's good. I can't believe it, there's no way I'm going back anywhere else. This is it. We're climbing here [at the moment] and I used to bloody surround my life [with competitive climbing]... My life was dedicated to this—just prove my worth through climbing, and excel and excel and excel. Now I'll still jump on a climb, but I've got nothing to prove. And eventually, I'm not going to climb any more. I tell people I can stop climbing right now because there's a priority that needs addressing and that's what I do. I could walk away right now from all that. But I'm still with people, I love to be with people. I need to talk to people; I need to talk to them about this. That's me for the rest of my life. How good is it to know what's going to happen for the rest of my life! Not thinking how I'm going to make money or... Yeah that's what I'm going to do. It's exciting.

[74] The other day I was walking around and this is my little folder. I've got my little flyers. I go past the houses and put them in letterboxes. Then I've got sticky tape and put these things [flyers] on. Yeah, when you're actually doing work for this, you can't help but to bloody smile all the time. It's healthy, I'm going for a walk. That's all I need to do. I don't need to go to the gym and get my six-pack and all that crap. I'm just going for a walk to put these flyers on. I've got a thousand stickers coming, you know, and I'm going to flood this place here, and then I'll expand, I'll keep going to Brisbane. I'll get caught probably and end up in police somewhere at some point. But then I can talk to the police people about this! [The WTM doesn't advocate breaking the law.] Yeah, that's me. Okay, I'll shut up now. But yeah, it's so good; it's so good. That 'Beach' Video just destroyed the old way of living. That's the real me, finally there's some realness to my bloody life—it's so good, so good.

Nikola Tsivoglou's Affirmation

Nikola Tsivoglou, Melbourne, Australia, April 2024

[75] **Nikola Tsivoglou** (from the Melbourne (Australia) WTM Centre): I just wanted to say that lately I've been in a peak Mexican Standoff state. I've just been falling back into the potholes of thinking in that old way—in that 'old world' thinking, feeling, behaving in that way—just feeling things are unjust and what-not, and sitting in that pain.

[76] It's funny, when I watched Jeremy's *Sermon On The Beach* video it just brought me back to when I first got this information. It was something that I was looking for my whole life and it was just so exciting to finally have the answers. It's so crazy that this is what everyone's been waiting for for so long, and yet you just somehow go back into your old way of doing things. It's a very slippery slope and you don't even realise that it's happening. But Jeremy's video just brought me back to the start of finding these answers and just how exciting it is!

[77] I really loved when Jeremy mentioned about St Paul and he said, "What if I just…?" I remember that was the line that I used when I first felt all this relief and the burden lift off my chest.

[78] These are tears of relief right now because…sorry. Honestly, I just feel excited! I've felt heavy for so long. And Jeremy's beach video, I don't know, I just feel lighter; I feel excited! What if I just fucking let go of everything!? Yeah, I'm just feeling that excitement again and that relief and…yeah, this is fucking crazy! This is fucking beautiful! I feel so lucky to have this, honestly. And I just wish everyone, everyone in this world could have it.

[79] **Tony Gowing**: Well how special was that, listening to Michael, Ales and Nikola—just so special!

[80] Now again, I absolutely encourage everyone to read the elaborated transcripts of Jeremy's *Sermon On The Beach* and Michael's, Ales's and Nikola's affirmations.

[81] Thanks heaps everyone for watching!

Postscript 1: The importance of Jeremy's *Sermon On The Beach*, by Tony Gowing

[82] This video that Jeremy Griffith made while sitting at the beach under a banksia tree is actually one of the most important presentations in human history!

[83] As Jeremy explains in paragraph 614 of *FREEDOM*, there have really only been four main events in recorded human history—four main events that helped the human race cope with the agony and horror of the human condition. Firstly, when the prophet Abraham was sound enough to think truthfully, such that 'The Lord [the truth of Integrative Meaning] appeared to him and said, "I am God Almighty"' (Bible, Gen. 17:1), and 'I am your shield' (Gen. 15:1), and as a result he introduced monotheism, the worship of the one *true* God (of Integrative Meaning), to replace the mad multiplicity of *false* gods, a god of war, a god of love, a god for eating our porridge, etc, etc. Then, secondly, when Moses, in his speech in Exodus 20 in the Bible, gave humanity the most effective form of Imposed Discipline for containing the ever-increasing levels of upset in humans through his Ten Commandments. Then, thirdly, when Christ gave humanity the soundest and thus most effective corruption-and-denial-countering religion through which the upset human race could live by deferring to the embodiment of idealism in him. And now, fourthly, the introduction of the Transformed Way of Living to cope with and solve forever our corrupted condition by living in support of the actual redeeming and rehabilitating biological understanding of our corrupted condition.

[84] Importantly, as Jeremy describes earlier in paragraph 44 of this elaborated transcript of his *Sermon On The Beach*, when Christ was

introducing his system of relief from humans' corrupted condition of having them live in support of his soundness and truth, everywhere he went he was running into the Mexican Standoff, where people's extreme attachment to their artificial angry, egocentric and alienated defensive ways of coping with their corrupted condition was blocking them from seeing how they could let all that go and instead live through their support of Christ's soundness and truth—with even his disciples complaining that **'This is a hard teaching. Who can accept it?'** (John 6:60). Eventually, seemingly out of frustration with the Mexican Standoff resistance he was encountering, Christ gave his central, seminal <u>Sermon on the Mount</u> speech, where, in Matthew 5:4-14, he essentially said, "Come to me all you who are suffering and I will give you rest." Basically, he was initially bypassing all those who were excessively locked onto their denial-addicted and power and glory defensive structure and as a result weren't wanting to admit they suffered from anything, and instead building a human-condition-relieved world through those who were the most open to the magnificence of the human-condition-relieved way of living he was offering.

[85] In the case of our <u>Transformed Way of Living</u>, in the analysis of where this human-race-saving fourth great event in human history is up to, it similarly has been blocked by the Mexican Standoff, of people's attachment to their old artificial denial and power and glory defensive structure. Sure we have been running into the problem of the Deaf Effect, and into the problem of the difficulty of not differentiating between the deluded and baseless and *false* human-condition-escaping offerings of pseudo idealistic New Age/PC/Woke charlatans and Jeremy's human-condition-confronting-and-solving *real* human-condition-free offering, but the actual problem, the actual

stalling point, has been the Mexican Standoff, because that is what is stopping those who can appreciate that the human condition has been solved from showing the rest of the world the magnificent life that breakthrough makes possible. So, at some point, just as what happened with Christianity, that Mexican Standoff situation had to be called out, stood up, exposed and put an end to, and that's what Jeremy's <u>Sermon On The Beach</u> does.

[86] <u>So, following Abraham's ability to think extremely truthfully, to 'chat with God', and create monotheism, we can clearly see that the three great speeches in human history are firstly, Moses's teaching of the Ten Commandments, then Christ's Sermon on the Mount to bust through the Mexican Standoff that his great relief system was up against, and now Jeremy's Sermon On The Beach to bust through the Mexican Standoff that his ultimate relief system has been up against!</u>

[87] <u>So that is how INCREDIBLY, IMMENSELY IMPORTANT Jeremy's *Sermon On The Beach* is!</u>

Postscript 2: Further inspiration from Ales Flisar & others

[88] This Postscript consists of some very informative discussion (edited for clarity) led by Ales Flisar that took place in our WTM Facebook Group in early June 2024.

[89] **Ales Flisar** (from the Sunshine Coast (Australia) WTM Centre): Yes, the way to be effectively free of the human condition is to live in support of the redeeming, reconciling and healing biological understanding of our corrupted human condition that the human race now, at last, has.

[90] Regardless of what we are doing, at any given moment, actually every second of every day, knowing that humanity's great, heroic, upsetting battle to find that redeeming understanding of our corrupted condition is finished with, is over, means there is no longer any reason to keep living out that battle. And that puts us in the position to take up the <u>Transformed Way of Living</u> where instead of *selfishly* focusing on denying our corrupted condition and trying to win power, fame, fortune and glory to artificially validate our worth, we *selflessly* focus on supporting and promoting the all-redeeming, all-reconciling and all-healing understanding of the human condition that we now have—which is a change of focus for us humans from living selfishly to living selflessly that totally fixes the world. It is, as Sir James Darling said, the **'serious and urgent' 'change of heart' 'decision'** needed for **'saving the world'**. [See the context of Sir James Darling's words earlier in paragraphs 27 and 52.] Yes, the wonder and magnificence of this change of focus is that it ends humans' selfish behaviour and by so doing fixes the world.

[91] <u>The benefit for us personally</u> is that, <u>firstly</u>, as Jeremy writes in *FREEDOM*, we are **'finally being genuinely aligned with the truth** [of Integrative Meaning] **and actually participating in the magic true world** [of our original all-sensitive and all-loving instinctive self or soul]' (par. 1239). In essence, we at last become part of the solution not the problem,

which is such an enormously joyful relief! Secondly, now that our corrupted condition is totally defended and understood at a fundamental level, we no longer have to be preoccupied every moment of the day trying to artificially and superficially make ourselves feel that we are good and not bad by denying the truth of our corrupted condition, and by trying to prove our worth by winning as much power, fame, fortune and glory as we can. The effect is that we legitimately and responsibly no longer have to focus on ourselves, and instead are free to focus on supporting and promoting the reconciling, redeeming and healing understanding of the human condition. And this legitimate and necessary mind-switch from focusing on ourselves to focusing away from self onto supporting the human-race-saving understanding of the human condition—which is what the Transformed Way of Living is all about practising—is so incredibly relieving it is almost indescribable!

[92] For me personally, this aspect of not having to focus on myself, in fact, responsibly *avoiding* focusing on myself where I am preoccupied denying the truth of my corrupted condition and winning as many accolades as I can, and instead focusing away from myself onto practically supporting the redeeming understanding of the human condition, is an absolutely all-relieving and all-exciting life-changer! And because it brings such incredible relief to my life I can see that it can bring incredible relief to everyone's life! As Jeremy writes in *FREEDOM* about the effects of such relief, 'It is this fact that there is no longer any reason to keep living out the battle to champion the ego that has the potential now to change the world so rapidly from one of conflict and suffering to a world of peace and happiness' (par. 1206). In summary, the less I think about myself, the better off the world is!

[93] **Stefan Rössler** (from the Austria WTM Centre): I really love your transformed new world thinking, Ales! And this would actually make a great poster to put on my wall: "The less I think about myself, the better off the world is, and it's completely legitimate, and it makes me feel so good!"

[94] **Ales**: Yes, I will write "The less I think about myself, the better off the world is!" on my wall, my car dashboard, the back of my phone, the front of my phone, on my toothbrush, on my bike, my work locker, on the top of my lunch box and water bottle. A constant reminder to my resigned mind that it is all over, the battle is finished with, and all I really need to do is to support this information.

[95] **John Biggs** (WTM founding member from the Sydney WTM Centre): Love every word of your post Ales!! It's absolutely inspiring how clearly you see the redundancy of our resigned strategies, and how embracing the Transformed Way of Living is the golden path home for the whole human race!!!

[96] **Ales**: Wouldn't want to be anywhere else John! Resigned strategies were full time artificial preoccupations for me that I can now understand were never going to make me happy, so to be able to be relieved of those preoccupations is the greatest relief and joy. We don't have to struggle for answers anymore, or seek validation anymore. So much time was spent on doing those things, which we now don't have to do, and indeed shouldn't do anymore; how freeing is that! All our questions have been answered through first-principle-based biological explanation, and our worth has been established at the fundamental level through actual redeeming understanding of our corrupted condition, so all our preoccupations with trying to understand ourselves and prove our worth changes to preoccupation with 'How can I spread this understanding and the Transformed Way of Living that it makes possible'. And whether our contribution to this task is small or big doesn't really matter; what matters is to have this information close to our hearts and our love of it in our minds every minute of every day—because this information is what saves us from bewilderment about the world, dread of our corrupted condition, and the constant futile search for happiness through our old human-condition-denying defensive structure. So let's go, let's get this human-race-saving Transformed Way of Living out to the world because it's now the only meaningful and responsible way to live.

[97] **Jack Soden** (from the Bolton (England) WTM Centre): Yes, my own and the world's condition is extremely upset, as it should be after 2 million years without understanding of the human condition. But now that we have this bridging instinct vs intellect explanation of the human condition which explains and defends that upset state, we have the solution to all of our own and the world's problems! The solution is to wrap both arms around this explanation (once we have verified that it is true) and to leave all our worrying and preoccupation with our now-explained-and-defended upset condition behind forever. Yes, it's that simple and straightforward! This understanding will do all the heavy lifting for us in terms of fixing the world and psychologically healing humanity, we just have to get right behind it and offer it all our support. As Jeremy said in his recommended mantra for the world now, 'If we look after this information it, in turn, will look after each of us and the world' (par. 1191 of *FREEDOM*).

[98] **Kevin Ryan** (from Dublin, Ireland): Yes, Albert Einstein famously said, 'The only way to achieve peace is through understanding', and since we now finally have the redeeming 'understanding' of the human condition that is 'the only way to achieve peace', what do we do with it?? WHAT WE DO WITH IT IS WE SUPPORT IT LIKE THERE IS NO TOMORROW!! And there actually won't be any tomorrow for us humans if we don't support it!!

[99] **Further comment from Jeremy Griffith**: In paragraph 53 of my booklet *Death by Dogma*, I describe how, "in 610 AD, which was 600 years of increased upset since Christ created Christianity, especially the increased upset in the cauldron of the Middle East", the simple solution to extreme upset of 'absolute submission to a unique and personal god, Allah' (*Macquarie Dictionary*) led to the incredibly rapid spread of Islam in the world. Well, the solution to the now utterly extreme levels of upset everywhere in the world of living in support

of the actual redeeming biological understanding of the human condition is infinitely more profound than deferring to a deity, and infinitely more straightforward to understand and appreciate, and thus infinitely more simple to take up—so it makes sense that this new human-condition-understood-based Transformed Way of Living is going to catch on and spread and create immensely excited relief <u>absolutely incredibly fast</u>—which is why I wrote in *FREEDOM* that since **'there is no longer any reason to keep living out the battle to champion the ego'** there is the **<u>'potential now to change the world so rapidly from one of conflict and suffering to a world of peace and happiness'</u>** (par. 1206). Again, as Teilhard de Chardin predicted, **'The Truth has to only appear once…for it to be impossible for anything ever to prevent it from spreading universally and setting everything ablaze'**!

[100] So yes, let's go, let's join Tony Gowing on the sunshine highway to the world in sunshine, let's get out of this hell hole of the human condition with all its horrendous suffering right away *because at last we really and truly can!*

Postscript 3: Jeremy's description of how 'falling in love' evidences the magnificence of a human-condition-free life

[101] We at the Sydney WTM Centre want to include this further illustration by Jeremy Griffith of how fabulous humanity's future is going to be.

[102] Firstly, we want to share this wonderful photo of Ales Flisar from WTM Sunshine Coast, Australia, who has been such an inspiration to us all with his Transformed State and his pure and

unwavering response to the flawless logic Jeremy has laid out for us—which you can see Ales talk about earlier in his awesome Affirmation (pars 64-74). Ales is spreading our life-and-world-saving good news to his whole community (and beyond!) by systematically mailbox-dropping leaflets, handing them out in public places, putting them up everywhere—and (as pictured earlier above par. 64) by driving around in his van that he has completely covered with pictures boldly publicising our explanation of the human condition, which Jeremy is so impressed with he said, **'Ales's van has to have a front-row spot one day in the WTM's 'hall of fame'!'**

[103] The reason we want to share this photo of Ales is because his face looks SO free!! As Jeremy said when he saw it, **'Yep, the face of the future human race! Serenity! Our father, who art in heaven, Ales's face be thy name, thy peacefulness come, thy happiness will be on Earth as it is in heaven!'**

[104] Inspired by this talk of the human-condition-free, heavenly state, Jeremy then gave this amazing description of how indescribably wonderful it is going to be living free of the human condition.

[105] He began by saying he saw this very short YouTube clip that he would like us to watch—www.wtmsources.com/304. It's from the 2012 film *The Lucky One* about a war veteran who walked all the way across the United States from Colorado to Louisiana to see a woman he had fallen in love with after seeing a photograph of her while he was serving in Iraq.

[106] In explaining the significance of this video clip, Jeremy wrote the following:

[107] "In paragraphs 786-790 of *FREEDOM* I describe how men and women could 'fall in love', let go of reality and dream of living in a human-condition-free, cooperative, selfless and loving ideal world; basically dream of being transported to how it once was before the human condition emerged—and to how it can be again now that we have solved the human condition—which is the heavenly state

that the *Lord's Prayer* looks forward to the arrival of, and which the freedom in Ales's face shows he is beginning to experience.

[108] As I wrote in *FREEDOM*, the lyrics of the song *Somewhere* from the 1956 blockbuster musical and film *West Side Story* provide a wonderful description of this dream of the heavenly, human-condition-solved state of true togetherness that humans allow themselves to be transported to when they fall in love: **'Somewhere / We'll find a new way of living / We'll find a way of forgiving / Somewhere // There's a place for us / A time and place for us / Hold my hand and we're halfway there / Hold my hand and I'll take you there / Somehow / Some day / Somewhere!'** The lyrics of the classic 1935 song *Cheek to Cheek* similarly describe just how intoxicatingly fabulous it is living free of the human condition, which is a feeling we are allowing ourselves to access when we fall in love: **'Heaven, I'm in heaven / And my heart beats so that I can hardly speak / And I seem to find the happiness I seek / When we're out together dancing cheek to cheek.'** Yes, absolutely, **'my heart beats so that I can hardly speak'** is how wonderful the ability to be free of the human condition is!

[109] Tragically having to live in denial of our corrupted human condition while we couldn't explain it meant we were also living in denial of our goal and dream of freeing ourselves from that corrupted state—if we are not corrupted, as we have had to delude ourselves that we are not, then there is no corrupted state to have to free ourselves from. Nevertheless, despite this tragic denial and delusion that we don't suffer from a corrupted condition that we have to free ourselves from, we have always *subconsciously* known

of that goal of freeing ourselves from our horrifically soul-destroyed condition, and it is that dream of **'somehow, some day, somewhere'** being able to live free of our corrupted condition that we are allowing ourselves to dream of when we fall in love. <u>So the incredible relief and indescribable happiness we feel when we fall in love is how wonderful being able to live free of the human condition is</u>, which the freedom and peace in Ales's face is showing, and it is the **'my heart beats so that I can hardly speak'**-experience that enabled the war veteran to easily walk all the way across America to be with the girl he has fallen in love with!

[110] <u>Imagine that, a heart beating so strongly that we can hardly speak, all day every day in everybody and towards everybody else and towards all of life! Falling in love is an incredibly powerful force that we can all experience, and through experiencing it we can all know how beyond description for us utterly exhausted, heroic humans living in that human-condition-free state will be! So let's go; sign me up Tony Gowing and Ales Flisar, let's leave this hell hole of the human condition and all join the transformed sunshine army on the sunshine highway to the world in sunshine!!!"</u>

Postscript 4: WTM's response to concern about Jeremy's shirtless 'sermon' on the beach

[111] This final Postscript is the WTM's response (which also appears as WTM FAQ 2.6) to the following comment that was made shortly after *Sermon On The Beach* was published on our website in June 2024: **'Is it not inappropriate to call biologist Jeremy Griffith's 'Sermon On The Beach' presentation a 'sermon', and isn't Jeremy being pretentious holding forth, shirtless, sunglasses and in an ideal tropical setting?'**

[112] Firstly, as WTM founding member Tony Gowing says in this Introduction to *Sermon On The Beach*, **'people aren't calling it a "Sermon" because it's religious, but because, like Christ's "Sermon on the Mount"—which was his great call to action, his great selling of the wonders of Christianity's ability to relieve humans of their distressed human condition—Jeremy's presentation is his great call to action to actually end the agony of the human condition that the finding of the redeeming and rehabilitating biological explanation of the human condition finally makes possible.'**

[113] In terms of Jeremy's presentation, it was a natural, spontaneous and inspired talk rather than something confected and pretentious. He had been thinking about the idea of people accepting the logic that the ever increasing levels of upset that results from searching for understanding will eventually lead to terminal levels of psychosis, which is the end play state happening in the world right now, and how that realisation can help people let go of their Mexican Standoff attachment to their egocentric, resigned competitive way of living and by so doing discover the relief of taking up the Transformed Way of Living. The reason for recording the speech on the beach is not in any way pretentious, Jeremy just happened to be there when all this thinking became very clear to him and so he decided to make the recording there and then. Jeremy's job is to think, and think about the most important of all subjects of the human condition that no one else is able to think effectively about

because they are all either living in a resigned state of denial of the issue of the human condition, or unable to overcome that all-dominating denial. Jeremy is our only hope of saving the human race from extinction from terminal alienation/psychosis, and he is fully aware of this responsibility. If you watch the video carefully you can see how motivated and focused he is, without there being any overtones of pretention. You can see him thinking very intensely about what he urgently wants to communicate, seemingly almost unaware that he is shirtless. Indeed, if he was concerned about his appearance at his age of 78 he would have likely embellished his appearance with clothes, make-up, hairstyling, etc. And as you can clearly see on the cover of this *Sermon On The Beach* booklet, the glasses hanging around his neck are not sunglasses, but the reading and long distance spectacles he, for practicality's sake, always wears around his neck. Jeremy has never worn sunglasses or a beard or tattoos or any embellishment for that matter; he is just a conscious thinking, thoughtful, considerate, loving mind walking around in the world seeking to make sense of it—as he has been able to reveal all humans really are at heart. Also, the beach is not some luxurious tropical island setting, it is simply one of the beaches near where he lives and drives to for some relief from the intensity of his work. On the east coast of Australia where Jeremy lives everyone goes to the beach—it has even been suggested that 'going to the beach' is the basis of Australia's egalitarianism, because on the beach without their shirts on, everyone, from the Prime Minister down, is equal! So, the whole presentation is unpretentious, not pretentious.

[114] It is natural for people to project the realities of the world, and even of their own insecurities, onto situations—in this case think that 'If someone, or even I, did what Jeremy is doing, holding forth without a shirt on, it would be some sort of ego trip'—but if you look more closely at what is happening, as has just been done, the evidence is that Jeremy is not on some pretentious ego trip, but is being completely unpretentious, unself-conscious and natural.

[115] And you can tell how accurate Jeremy's thinking about the need for his talk is, and therefore why he wanted to capture it straight away, by the effect the talk has had on people. For example, Ales Flisar's and Nikola Tsivoglou's affirmations included at the end of the talk express how it helped them enormously. Jeremy's mind was reading the play, the need for that talk, extremely accurately!

[116] Jeremy's honesty, sensitivity, lack of pretence, absence of ego-centricity have been the defining traits of his whole life. The obvious truth is that he couldn't have made any real progress towards all the amazing truthful insights he has reached if his mind was unsoundly motivated by egocentricity or self-aggrandisement or any of those insecure behaviours. As Christ said when he was accused of being a deluded, falsely-motivated, **'prince of demons'** (Mark 3:22) person, **'a bad tree cannot bear good fruit'** (Matt. 7:18) and **'How can Satan drive out Satan?'** (Mark 3:23). Those who work and interact with Jeremy all the time have never experienced him being interested in self-aggrandisement. All Jeremy is interested in is getting the truth up and stopping the suffering on Earth. In fact, one of the problems the WTM has with people when they meet Jeremy is they are expecting somebody who puts on the airs and pretences of academic experts and new age gurus, etc. What they encounter is someone who is completely natural, almost childlike in his lack of any pretence.

[117] Certainly, scepticism is a legitimate and responsible initial attitude to the analysis of the human condition in Jeremy's work, which is why it is important to assess its merits or otherwise on the accountability of his explanations, and, in terms of his behaviour, on fair and careful analysis of that.

www.ingramcontent.com/pod-product-compliance
Lightning Source LLC
Chambersburg PA
CBHW051504270326
41933CB00021BA/3459